What
Matter
Feels

What Matter Feels

Consciousness, Energy and Physics
(How Science can Explain Minds)

This treatise is a working document. It outlines a framework for studying the relationship between mind and matter using knowledge from art, philosophy and science. The aim is to encourage discussion and stimulate research between diverse disciplines to create a more unified understanding of the world and our place in it. Most significantly, it defines new psychological properties of matter and provides methods for experimentally testing hypotheses that may explain the neurobiology of pleasure and pain, the evolution of life, aesthetic value and consciousness.

Robert Pepperell

Version 1.0

IRM
LONDON

Pepperell.art

WHAT MATTER FEELS
Consciousness, Energy and Physics

PEACE
The equalisation of all forces

PLEASURE
Rewards useful behaviour

ACTION
True contradiction

Visit **www.pepperell.art** for a range of high-quality art posters that illustrate and celebrate the ideas presented in *What Matter Feels*.

Cover artwork and design by Robert Pepperell.
All illustrations, diagrams and artworks are by Robert Pepperell unless otherwise stated.

Produced by Pepperell.art (for IRM Editions).
Paperback edition: ISBN 978-1-0685232-0-5.
Proofreading and copy editing: Mark Thompson.

Published by:
IRM (Interdisciplinary Research Media) Editions,
45 Fitzroy Street,
London, W1T 6EB, United Kingdom.
www.irm-editions.com
For all inquiries email: info@irm-editions.com
(No postal inquiries accepted)

IRM
EDITIONS

What Matter Feels is dedicated with gratitude to Ruth, Daisy and Emily and to my late father, Derrick.

Contents

Contents

Acknowledgements

I have lost track of the many people who contributed over many years to the development of the framework outlined in this treatise. I am grateful to them all, especially those who listened with patience (and, I suspect, forbearance) as I struggled to present these ideas as a coherent and useful thesis.

Special thanks go to Nathaniel Barrett and Clive Cazeaux, who engaged in fruitful discussions and gave critical feedback on some philosophical aspects of the text (although they are not responsible for any philosophical deficits it may now contain), to Alan Dix for his questioning and enthusiasm and to Alistair Burleigh for showing how to combine the technical and the imaginative.

Thanks also go to the editors and reviewers of academic journals from fields such as physics, psychology, neuroscience and philosophy who commented on papers I submitted containing these ideas as they germinated. That so few of those submissions were accepted is testament to the admirable rigour of the peer review process—which rightly repulsed some of my early attempts to articulate this framework—but also to the instinctive resistance of contemporary academia to pan-disciplinary approaches. I am hopeful that future generations of researchers will meet less resistance.

Finally, I wish to acknowledge the contribution made by Mark Thompson. As a highly proficient copy editor he corrected countless grammatical and typographical errors along the way. But more significantly, as an open-minded and widely informed scientist he continuously challenged me to make the ideas in the book clearer and more consistent. If there is any scientific value in the ideas contained herein, then he must take a good deal of the credit.

And only because you are a part of the world you see within yourself a part of that which the world sees in itself.

Gustav Fechner, *The Little Book of Life After Death,* 1904

Preface

What Matter Feels is a bold—perhaps overly ambitious—attempt to outline a conceptual framework that tackles one of the deepest intellectual problems we face: how to explain the relationship between matter and mind. Recent years have seen numerous attempts to solve this problem using existing frameworks from science and philosophy, whether they be computational, neurobiological, mathematical, phenomenological or quantum physical.

While all of these have made valuable contributions, it is widely accepted that none has fundamentally unified our knowledge about the nature of matter from physics with that we have about the nature of mind from psychology. None of these approaches allows us to experimentally study the mental states of physical systems using principles that might explain how those systems have mental states at all. Had we solved what David Chalmers in 1994 called "the hard problem of consciousness", we would not still be arguing about it.

The framework outlined here offers an opportunity to address these challenges experimentally. It sets out a model of mind and its place in nature that draws on a wide range of current scientific and philosophical knowledge but also on the history of that knowledge, and especially on certain ideas that were prominent in the nineteenth century but are now largely overlooked.

At the core of the proposed framework is a conjecture that Isaac Newton hinted at in his foundational work, the *Principia Mathematica*, as discussed in part I of this treatise. The conjecture is that for the purposes of scientific study we should consider matter to have, in addition to properties such as mass and charge, a general property called Experience. I use the word 'Experience', capitalised, as a noun in a technical sense that is distinct from—but related to—the various senses in which we use it as a noun or verb colloquially.

Experience in this technical sense is a psychological property that belongs to a portion of matter from its own point of view. It cannot be observed from the 'outside in'. Yet as I will show, it can be measured. There is a precedent for an unobservable yet measurable property of nature, which is force. A force cannot be seen. But it can be measured by the changes in motion it produces in a system it acts upon. As will be explained, the psychological property of Experience is measured by the changes in energy a system undergoes when it is subjected to work (as defined in physics). Besides the *general* psychological property of Experience, I also define *specific* psychological properties that can be attributed to matter such as Will, Volition and Conflict and show how these can be measured.

The phrase "from its own point of view" above is of utmost importance and highlights one of most conceptually challenging aspects of *What Matter Feels*. For the purposes of theoretical and experimental investigation, I ask us to entertain the possibility that *all* material systems can have an intrinsic perspective, i.e., there can be "something it is like" subjectively to be a material system (to use a phrase given prominence by the philosopher Thomas Nagel). In saying this, I am following in the wake of earlier scientists such as Gustav Fechner and Ernst Mach in suggesting, as the biologist Ernst Haeckel put it, that matter may be "sensitive": that matter, as claimed in the title of this treatise, "feels". I take this suggestion seriously and literally for the purposes of scientific study. It may, of course, prove to be wrong or require revision in light of experiment.

I stress that I am *not* saying—as some contemporary panpsychists do—that all matter is *conscious* at some fundamental level, consciousness being defined here as a condition in which a system is aware of itself and its environment. The question of how a given system might become conscious, or even self-conscious, rather than just having a capacity for Experience is addressed in this treatise. Using evidence from contemporary neuroscience, I describe how nervous systems and brains undergo conscious Experiences when they are organised and behave in certain ways. Conscious and self-conscious Experience, I try to show, are special cases of the general property of Experience.

Many people will balk at the idea that matter can feel or Experience or have sentience of any kind. Yet generations of chemists and physicists have routinely referred to the feelings and preferences of the material systems they study. The earliest example I can find in the modern literature is the physicist James Clerk Maxwell in the 1880s talking about the "experience" of material bodies acted on by forces. But earlier natural philosophers such as René Descartes, Isaac Newton and Michael Faraday also talked about the way material bodies "suffer" or "endeavour" during interactions. Chemists still talk of 'happy' and 'unhappy' reactions and physicists of systems that 'want' to do some things and not others.

Whenever I have asked chemists or physicists what they mean by talking in this way, I invariably get the same response. They say that they are speaking metaphorically; they do not mean to impute psychological states to atoms, liquids or mechanical gears. Matter self-evidently does not feel, they claim. Thinking a bit more carefully, though, how secure is this claim? Has it been tested experimentally? I am not aware that it has. This treatise aims to provide reasons why it should be tested and to suggest how it might be.

What Matter Feels has some unusual features that are both strengths and weaknesses. For one thing it is promiscuously multidisciplinary, drawing on knowledge and methods from the arts, humanities and sciences. The strength here is that the resulting framework benefits from a broader range of perspectives than a monodisciplinary approach could provide. In my view, the nature of mind and its place in the world are topics that do not respect academic boundaries, and they are certainly not the exclusive concern of philosophy or science. The weakness of the present approach is that the depth and subtlety of the ideas contributed by some disciplines may not be given their full due.

The treatise is also quite short, especially in relation to its scope, and this limits the space for detailed and sustained argument in support of its main claims. But its brevity is also a strength, not just because of the reduced demands it makes on the reader's time and on the earth's resources but also because it is only as long as it needs to be to fulfil its main purpose. It is not intended to convince by argument but rather to outline and illustrate a conceptual framework within which experiments can be conducted to test hypotheses and make predictions. Judgement on the contents of this treatise, I ask, should be withheld pending the outcome of those experiments.

Finally, although this treatise is published it is not finalised. Rather, it is a working document that is designed to encourage discussion and stimulate research. Hence, it is designated 'version 1.0'. One of the most valuable lessons I have learnt from my involvement in science is how often rational expectations are confounded by experimental results. When the experiments proposed here begin and data is collected, the most likely outcome—based on my previous experience—will be the need for profound revision of what in the absence of that data seemed to be reasonable conjectures.

Robert Pepperell, October 2024

Part I

I.1 The problem

The problem is dizzyingly difficult. It concerns that which is most real to us yet lacks substance. It concerns that which we share publicly yet remains most private to us. It concerns the fact that we can think, imagine, remember, believe, reason, doubt, learn, forget, love, know and dream, and express those things to others who seem to be able to do the same. We can behold the world in detail, colour and depth. As you read, the words written here are spoken somewhere inside you.

How? Why? We could have evolved as automata, as zombies, programmed to sense without sentience. But not only are we sentient, we are aware of being so and can ask why. We are self-made agents that feel our way through space and time, tracking the objects we desire. Volition is a primordial mover, and our actions are governed by basic emotions as the philosopher Jeremy Bentham observed:

> Nature has placed mankind under the governance of two sovereign masters, pain and pleasure. It is for them alone to point out what we ought to do, as well as to determine what we shall do.[1]

---

[1] Bentham (1789).

Or another philosopher, John Locke:

> Pleasure and pain and that which causes them—good and evil—are the hinges on which our passions turn.[2]

All these phenomena are expressions of the fact that we feel and know the world and ourselves. In short, we have minds.

Philosophers such as Immanuel Kant have argued that our minds shape our knowledge of the material world—a world that we can never know directly—by structuring the data we get from our senses. Other philosophers, such as Bishop Berkeley, argued that we are *nothing but* minds and there is no material world beyond our minds to know. Nevertheless, we still instinctively believe in an external world made mostly of numb, dumb and thoughtless matter—the very same stuff that we and our minds are supposedly made of. The main problem addressed in this treatise is how to make sense of this. How does mindful matter like us exist in a mostly mindless world? How does matter get its mind?[3]

---

[2] Locke (1690).

[3] A recent episode brought home to me the importance of these questions, and perhaps of the approach I have taken to addressing them. As I was coming to the end of writing this treatise, I woke one morning with an epiphany that I am first and foremost an *experiencing being* and that my experience is as much a part of the world as the objects around me. Moreover, I felt with certainty that my experience is an *intrinsic* property of my own being and that all things in nature must likewise experience themselves from their own intrinsic perspectives. I then found much the same idea expressed by Gustav Fechner in the quote preceding the preface. Whether this epiphany was the product of my subconscious mind justifying *post hoc* the ideas contained herein, I cannot say. But the episode left me feeling that I had intuited a deep truth about nature that is so obvious and so pervasive that it is hardly ever acknowledged. This treatise—I now realise in retrospect—is an attempt to articulate that intuition, which has been gathering in my mind for years, and to explore whether it does indeed contain any truth or, perhaps more saliently, any scientific value.

I.2   The complexity of the problem

The problem of how mind relates to matter is an ancient one that is still far from being solved. Paradoxically, we can scientifically explain more about the workings of the material world, which we access only indirectly through our senses, than we can about our minds, which we access directly. We understand a great deal about the physics, chemistry and biology of brains, for instance, and how certain neural activity correlates with certain mental states. But scientists and philosophers are still at a loss to explain how that activity might cause or give rise to our conscious experiences.

Perhaps, as some people think, our inability to solve this problem is due to the sheer complexity of the brain's organisation and behaviour. According to neuroscientist Anil Seth, the mystery of consciousness should gradually dissolve as our tools and methods for unravelling that complexity improve over time.[4] Nevertheless, the more we study the brain, the more neurobiological intricacy we encounter. It is becoming clear, for example, that individual neurons—once modelled as simple binary relays—are immensely sophisticated biochemical processing systems; the problem may deepen rather than dissolve as we gain more knowledge.

While the complexity of the brain is doubtless a major barrier to our understanding, I do not think of this as the main reason we struggle to explain why we have minds or how brains might produce them. Indeed, brain activity is in some ways quite simple when considered at a fundamental level. It consists of no more than *interactions between matter and energy*. The brain is made up of varieties of organic matter

---

[4] See Seth (2021). For a recent comprehensive and well-balanced overview of prominent approaches to the problem of explaining consciousness, see Kuhn (2024).

that interact with thermal, chemical and electrical energy that is generated by and regulated by metabolic reactions in cells. The energy produced in the brain is used to do physical work on its organic matter, exerting forces that move it in exquisitely choreographed ways to maintain its structure and carry out its functions.

In this respect, brain activity is no different from anything else that happens in nature. All natural processes can be described in terms of the transfer of energy between portions of matter: when the sun shines, its energy is transferred to organic matter on earth; when we eat, the energy stored in food (which is derived from the sun) is transferred to the matter in our bodies so we can move; when we kick a ball, we do physical work on the ball to give it kinetic energy using the power we get from the food we eat (which ultimately comes from the sun). However complex, however big or small, the processes described by physics, chemistry and biology—even psychology—depend on energy transfer in matter.[5]

The reason we struggle to explain the relationship between minds and brains is not just because of the formidable complexity of the biochemical processes involved. It is because of our inability to imagine how mental states can exist within material systems that, at a basic level, are doing no more than transferring energy between their parts—however complex the processes involved might be. As we watch the dance of energy and matter, we find no place for mind.

---

[5] In this treatise, I consider natural processes that entail interactions between energy and matter within the framework of classical mechanics—mainly Newtonian mechanics and classical thermodynamics—rather than quantum or relativistic mechanics, or for that matter any of the other branches of science that study motion and change. However, I am also aware that the proposals contained herein have implications for the study of natural processes at all scales of time and space. Some of these implications are touched on briefly in relation to quantum-scale processes, but it would require a much longer volume to begin to explore them in any depth.

I.3   Energy, matter and mind

The purpose of this treatise is to outline a framework that may help explain how minds are produced when energy and matter interact. If all natural processes consist of energy transfer within systems of matter, and if mentation is a natural process, then mind must be explicable in terms of energy transfer within material systems.[6] The physicist James Clerk Maxwell neatly set out this situation in 1877:

> All that we know about matter relates to the series of phenomena in which energy is transferred from one portion of matter to another, till in some part of the series our bodies are affected, and we become conscious of a sensation.[7]

One of the founders of modern neurophysiology, Charles Sherrington, was deeply puzzled by the problem of how minds could arise from energy transfer:

> Here we have of course to remember that in looking for mind as energy we are not looking for a form of energy then to translate into mind. Of that we have abundant instances already. Thus, radiant energy via nerve into seeing, or into heat-sensation, or pain. That would be to look merely for forms of energy which nerve can transmute through sense into the mental. What we look for is an energy which is mind.[8]

---

[6] Energy and matter are scientific concepts that gradually formed over several centuries and are still evolving today. For the purposes of this treatise, I use the standard definitions from classical mechanics in which energy is the capacity to do work and matter is that which has volume and mass.

[7] Maxwell (1877).

[8] Sherrington (1940). For further historical context on this point see James (1890), and in particular the discussion of "mind-stuff" theory in chapter VI. See also part III.7 of this treatise.

The framework outlined in this treatise for tackling the main problem, which includes a definition of Sherrington's "energy which is mind," is quite straightforward even though it may be difficult to understand at first. It is not entirely new, yet it has not been expressed so precisely or explicitly before. It rests on the counterintuitive conjecture that *Experience is a property of matter*. This conjecture was introduced in the preface and will be discussed in depth throughout this treatise, where many examples are given to clarify its meaning.

The framework is straightforward insofar as it simply adds another property, Experience, to the 65 that are currently listed on Wikipedia.com under 'Physical property', where it sits alphabetically between 'emission' and 'flow rate (mass)'. We can then use this property, which is defined in part II of this treatise, when scientifically studying the organisation and behaviour of material systems—both living and non-living—just like any of the other 65 properties.

The major difference between Experience and other established physical properties is that Experience is *psychological*, as noted in the preface. This means it refers to the *intrinsic* mental condition of a material object, which we cannot observe from an *extrinsic* or external perspective. I will explain exactly what this means and discuss illustrative examples, again in living and non-living systems.

Many people will instinctively reject the idea that matter can have psychological properties. But I will show that this conjecture can generate new hypotheses, testable predictions and possible explanations for phenomena such as pleasure and pain, mental causation and even consciousness itself. This is what counts in science; whether an idea makes instinctive or intuitive sense is far less important than whether it can be empirically tested and lead to better explanations of natural processes. To understand the proposed framework for solving the main problem, how it can be tested, and what it helps to explain we first need to set it in historical and philosophical context.

I.4   What nature is

Philosophers in the European tradition have tended to separate nature into material and mental realms. The material realm consists of inanimate objects and substances—various forms of matter—whereas the mental realm consists of thoughts, feelings and experiences—various forms of mind. Arguments have endured for millennia about how these realms can be divided and united. The result, broadly speaking, is a stalemate between two sets of entrenched and opposing views: materialism vs idealism and monism vs dualism.

Materialism, which underpins the standard scientific worldview, holds that everything that exists is a form of matter and that minds are products of the way material systems such as brains are organised and behave. Precisely how mental states are produced by, from or within material systems like brains, or how they come to be part of the material world at all, is unclear and remains a matter of debate.

Idealism, which is less favoured among scientists than philosophers, holds that everything that exists is a form of mind and that the material world is a product of the way the mind constructs our experience of reality. Nature *seems* to be made of solid material objects, as Dr Johnson showed by kicking a stone when attempting to refute the idealist philosophy of Bishop Berkeley.[9] But for idealists these objects are no more than ideas or impressions produced by our minds as they interpret sensory data from a material world of doubtful existence.

Monism, which can be either materialist or idealist in flavour, is the view that nature is made of only one thing or substance, in contrast to dualism, which holds that nature consists of two essentially different kinds of things or substances: mind and matter. Materialism is often taken to be a form of monism in that it assumes matter is the

---

[9] As recorded by Johnson's biographer James Boswell (1791).

fundamental 'stuff' and mind is somehow made of or caused by it. But idealists can also be monists when they say all that exists are ideas. Some dualist positions—like that of René Descartes, who held that the mind, our knowledge of which is indubitable, is indivisible while the material body, from which we gain knowledge that is dubitable, is divisible—seem to allow for interactions between mind and body.[10]

The framework outlined here does not adhere to any one of these conflicting philosophical positions on the grounds that none alone has yet proved suitable for addressing the main problem scientifically. Instead, I propose that any material system (portion of nature) should be treated *for the purposes of scientific study* as monistic, i.e., as something in which there is no inherent division between the material and the mental and, *at the same time*, as something that can be considered from two parallel points of view—one material and one mental.

For example, when scientifically studying a system such as a living cell, we should treat it as having both a material nature and a mental nature, neither of which can be isolated from the other since they are both part of the same nature. This nature can be considered from two different perspectives: the cell's material nature is evident when we observe it from our own point of view, while its mental nature is what the cell experiences from its own point of view. In short: *matter is how a system appears to an observer from an external perspective whereas mind is what the system experiences from its internal perspective.* This monist yet 'dual-aspect' approach will be discussed in more detail shortly, along with its historical precedents.

---

[10] Descartes' reputation for having introduced the split or rift between mind and body into modern philosophy is not entirely deserved. In fact, his view was more subtle and in some ways contradictory. In the synopsis of the *Meditations*, for example, he says, "The human mind is shown to be really distinct from the body, and, nevertheless, to be so closely conjoined therewith, as together to form, as it were, a unity" (Descartes, 1637/2006). For Descartes, it seems, the mind and body are distinct *and* unified, both two *and* one, as will be discussed further in part III.1.

I.5   Sensitive matter

A central question here, perhaps *the* question, is whether matter in general can have mental states and points of view. Do cells, liquids and bundles of chemicals—or more fundamentally, electrons or atomic nuclei—have feelings in themselves? Is matter, as many nineteenth century biologists thought, "sensitive"?[11] People today might naturally assume not.

But we have conclusive evidence that at least one kind of material system composed of cells, liquids and chemicals has mental states from its own perspective. Us. If you were to observe me or analyse my composition from *your* point of view, which is extrinsic to me, I would appear as a material system made of fats, proteins, water and sundry other compounds, all of which are in turn composed of atoms and subatomic particles such as electrons and nuclei. From *my* point of view, intrinsically, I am primarily an experiencing agent who exists as a bundle of mental states, just as you are from your point of view. The same object appears as a complex of material properties from one perspective and a complex of mental properties from another.[12]

This should strike you as strange: each of us is made of numb, dumb and thoughtless matter yet each of us can think and feel. And it's not that there's anything special about the matter in us that thinks and feels. The particles that make up the matter of our brains—the

---

[11] In *The Riddle of the Universe at the Close of the Nineteenth Century* (1901), the artist and biologist Ernst Haeckel favourably quotes Goethe's dictum: "Matter cannot exist and be operative without spirit, nor spirit without matter." For Haeckel, "sensitive substance" was a fundamental attribute of the world.

[12] Thus, when John Locke says we must choose "between two accounts of what a human being is: 1. It is a material thing that thinks 2. It is a material thing linked with a second thing that thinks," we in fact have a third choice: it is *both*. See Locke (1690). Note also that when I observe myself, I experience myself as both a material object and a mental agent. This issue will be further discussed in parts I.20 and III.6.

electrons and nuclei—are identical to those in stones and chocolate. Perhaps there is something special about the way the matter of the brain is organised?

We will return to that question later, but for now, this thought exercise teaches us that matter—at least the matter in our nervous systems—*can* think and feel. However, the question of whether *all* matter, not just that in nervous systems, can support mental states remains open. Here I will show that by treating matter as being endowed with mental properties for the purposes of scientific study we can address the central problem empirically.[13]

1. *Sensitive matter*, 2012. Indian ink on paper, 30 × 55 cm. To you, I appear as a bundle of chemicals and compounds whereas to myself I am a bundle of thoughts and feelings, just as you are to yourself.

---

[13] I am not advocating for a form of panpsychism which holds that "consciousness is a simple, fundamental and pervasive—perhaps ubiquitous—element of reality" as defined by Seager (2016), for reasons given elsewhere in this treatise.

I.6   Psychophysical parallelism

Perhaps the first person to set out a monist yet dual-aspect conception of mind and matter of the kind proposed here was the seventeenth century philosopher Baruch Spinoza, who argued that:

> ...thinking substance and extended substance are one and the same thing, which is now comprehended through this and through that attribute.[14]

Spinoza was a monist in the sense he held that only one substance exists. But he also held that this one substance has many aspects, only two of which (mind and matter) can be comprehended by us, the rest belonging to God. In the nineteenth century, Spinoza's view was to influence one of the founders of experimental psychology, Gustav Fechner, who spent much of his career developing a research method called 'psychophysics' that aimed to integrate psychology with the physical sciences. Similarly to Spinoza, Fechner held that:

> What will appear to you as your mind from the internal standpoint, where you yourself are this mind, will, on the other hand, appear from the outside point of view as the material basis of this mind.[15]

There are subtle but significant differences between Spinoza's and Fechner's monist yet dual-aspect views, which also overlap with ideas expressed by Gottfried Leibniz, who was active in the seventeenth and eighteenth centuries. These varying perspectives will be discussed further in part III of this treatise, but the most important point here is

---

[14] Spinoza (1677/1954). Like many of his time, Spinoza distinguished between the mind, which is unextended (locationless) in space, and matter that is extended in space.
[15] Fechner (1860/1966).

that during the late nineteenth and early twentieth centuries, the dual-aspect view was broadly accepted by many in the scientific community, not least by physicists such as Niels Bohr and Albert Einstein who knew it by the name 'psychophysical parallelism'.[16] In a letter of 1922 Einstein wrote:

> To avoid a collision between the different sorts of 'realities' that physics and psychology deal with, Spinoza or, resp., Fechner invented the doctrine of psychophysical parallelism which, quite frankly, fully satisfies me.[17]

One reason to take the doctrine of psychophysical parallelism seriously today is that it offers a useful alternative to the view now widely held by scientists and philosophers that physical events such as brain processes *cause, give rise to,* or *are the basis of* psychological events such as conscious experiences. The precise nature of the mechanisms causing psychological events to occur—whether they be biophysical, biochemical, neurophysiological or even quantum mechanical—are currently unknown, but many theories are being developed that aim to explain what the basic processes might be.[18] This prevalent model of the relationship between physical and psychological events, where the latter are mechanistically *caused by* the former, is illustrated schematically in figure 3. A major and recognised weakness of this model is

---

[16] For a detailed account of the development of these ideas, see Heidelberger (2004); on a related version of 'dual-aspect monism' developed by Wolfgang Pauli and Carl Jung, see Atmanspacher (2012); and on 'double–aspect' theory see Chalmers (2007).
[17] Published in Einstein (1922/2012).
[18] Seth and Bayne (2022) reviewed several current theories of what they call "the biological and physical basis of consciousness". In the absence of an explanation of this physical basis, researchers often refer only to the 'correlation' or 'association' between neural states and mental states. See also Francken et al. (2022) for a survey which showed that most academics in the field believe minds are *physically* explicable.

that it contains explanatory gaps between physical and psychological processes, as shown in the figure.

In contrast, the claim of psychophysical parallelism in its general form is that physical and psychological events—or what Einstein called "realities"—coincide but have no causal interaction. As we will see in part III.3, Leibniz conveyed this idea using the analogy of two clocks that run in perfect synchrony without physically influencing each other. Expressed in this way, there is a risk that psychophysical parallelism might be misconstrued as just another form of philosophical dualism, and as such would offer no new explanatory potential or scientific value. But I will show that when properly understood and applied, it *does* have explanatory potential and scientific value.

In the form of psychophysical parallelism adopted here and based largely on that proposed by Fechner, a scientist studying or explaining the behaviour of any material system—not only a nervous system—should treat it as a *single* entity that can be considered from *two* parallel perspectives.[19] The physical perspective is that which the scientist observes from an external viewpoint, objectively, while the psychological perspective is the internal viewpoint of the system that belongs to the system alone, subjectively. The latter cannot be observed directly by the scientist, even though its existence and behaviour can be inferred—at least hypothetically—from external observations.

To make this clearer we can draw a helpful analogy with the act of observing a full moon, as illustrated in figure 2. Just as there is a *single* moon that has *two* parallel sides—one observed and one inferred from observation—the form of parallelism adopted here treats the system for experimental and explanatory purposes as being *one* thing that can be observed or considered from *two* sides.

---

[19] As noted, the form of psychophysical parallelism developed in this treatise owes much to Fechner, but also draws on ideas from Spinoza and Leibniz as well as Isaac Newton and Ernst Mach, as will be discussed further in part III.

With respect to causation, when observed from its physical perspective the system's behaviour can be explained in terms of physical causes of the kind familiar to science (i.e., forces, charges, fields, etc.). Considered from its psychological perspective, on the other hand, I propose that the system's behaviour can be explained in terms of complementary or parallel *psychological* causes of a kind not yet familiar to science. As will be discussed in part II, these psychological causal agents include a property called Conflict—which is a psychological parallel of physical force—along with others such as Will and Volition. Definitions of these terms will be provided.

For experimental and explanatory purposes, then, we can treat any event as being causally determined while recognising that the specific cause can be considered from either its physical or psychological aspect. The parallelism and unity of these physical and psychological chains of causation is illustrated in figure 4, which shows the relative parsimony of this model and its lack of explanatory gaps.[20] In sum, the psychological chain of causation coincides with, and indeed *is identical to*, the physical chain of causation such that they constitute the *single cause* of the sequence of events being studied.

The main purpose of this treatise is to outline a scientific framework based on this form of psychophysical parallelism that enables us to study the psychological states of material systems by inference from observations of their physical states. The reason for doing so is to explain the psychophysical behaviour of material systems better than is possible with reference to physical appearances and causes alone. In this approach, we are not required to adopt any stance on the metaphysics of mind–matter relations; these will remain a matter for philosophical debate. But we can proceed with scientific investigation.

---

[20] Note that psychophysical parallelism as defined here does not refer merely to a correlation or association between brain states and mind states. Correlation and association tell us nothing about causation, whereas this form of parallelism does.

To be scientifically useful, however, the framework will need to develop something like a new 'physics of the mind'—that is, an extended form of the psychophysics developed by Fechner—which allows us to measure, experiment on, predict and causally explain the behaviour of psychological events with the same precision that we now measure, experiment on, predict and causally explain the behaviour of physical events. Doing this will enable us to put the discipline of psychology, with its concern for subjective qualities of mind, on a parallel experimental footing with the discipline of physics, with its concern for objective quantities of matter. How can this be achieved?

2. *Psychophysical parallelism.* This image illustrates the analogy introduced in the text between observation of a full moon and the experimental approach to mind–matter relations in psychophysical parallelism. The observer (left) can see the side of the moon that reflects light from the sun (by analogy, its physical side) but not its parallel dark side (by analogy, its psychological side), the existence and behaviour of which must be inferred from what can be observed. Using this conceptual framework, the system (the moon) can be treated for experimental purposes as being *one* system that may be considered from *two* aspects.

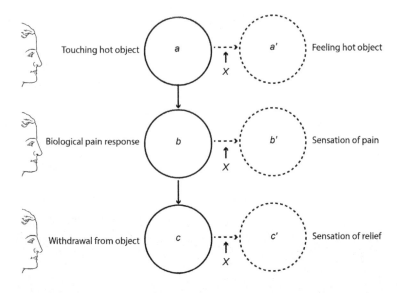

Touching hot object — *a* ⟶ *a'* — Feeling hot object

Biological pain response — *b* ⟶ *b'* — Sensation of pain

Withdrawal from object — *c* ⟶ *c'* — Sensation of relief

3. This figure is a schematic model of the view that physical events *cause, give rise to,* or *are the basis of* psychological events. The solid circles labelled *a*, *b* and *c* represent a sequence of physical events—causally connected by solid arrows—that occur when a person being studied by the scientist pictured on the left touches a hot object and quickly experiences a sensation of pain, prompting withdrawal of the hand to obtain relief. The physical aspects of the events can be observed by the scientist. Each physical event in the system supposedly causes—or is the basis of—a correlated and unobservable psychological event that occurs in the person, labelled *a'*, *b'* and *c'* in the dashed circles. Note that the causal link between the physical and psychological events, indicated by the dashed arrows, is assumed but cannot be explained by this model as the mechanisms are not understood. Hence, this model contains explanatory gaps, indicated by each **X**.

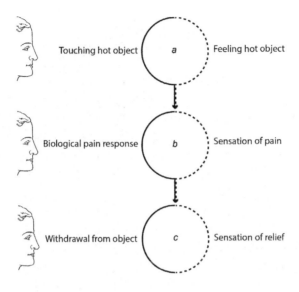

4. This figure is a schematic model of the same causal chain according to the form of psychophysical parallelism adopted here. As with the moon analogy, for experimental purposes we can consider the system being studied—in this case the person carrying out the act of touching a hot object—from two parallel aspects, one physical and one psychological, represented by the solid and dashed parts of each circle, respectively. The single causal chain in the sequence of events labelled *a*, *b* and *c* can be considered as both physical (indicated by the solid side of each arrow) *and* psychological (indicated by dashed side of each arrow) in nature. Each link causes a subsequent (observable) physical or (unobservable) psychological event, respectively. This causal chain, indicated by the conjoined arrows, explains the psychophysical events in full. Note this model is more parsimonious than the one in figure 3 and contains no explanatory gaps.

5. *Nature is exceedingly simple (Isaac Newton)*, 2020. Ferrofluid in petri dish, 6 × 6 cm. This artwork contains a geometrical pattern created by magnetic 'lines of force' (see part III.5) interacting with a colloidal liquid containing ferromagnetic nanoparticles. Isaac Newton realised that nature is driven by interactions between bodies (matter) and forces occurring at scales from the microscopic to the cosmic. It is the action of a force upon a portion of matter that causes the matter to be, as he wrote, "put out of its state of rest or motion".

I.7   Newton's active matter

We can begin to address the question posed at the end of the last section by considering the origins of modern physics. One of its founders, Isaac Newton, had a profound insight as a young man that was to develop into one of his most important contributions to mechanics and to physics in general.[21] His predecessors Galileo and Descartes had believed that material bodies, once moved, were kept in constant motion by internal 'forces' unless interrupted by contact with another body. But the nature of force and the role it played in motion were poorly understood; it remained, in Richard Westfall's words, a "treacherous" metaphysical concept.

Newton realised that force could be quantified by measuring the *change* in motion—what we now call acceleration—that it produced in a body when acting on it externally.[22] In later formalising this insight in his laws of motion, Newton did not tell us what force intrinsically *is*, but he did tell us how to *measure* it from an extrinsic perspective.[23]

Newton's realisation that a body or portion of matter must be acted upon by an external force if its motion is to change, and that this change can be quantified in a way that is proportional to the body's mass, lies at the foundations of classical mechanics—the study of forces and motion. That this applies to interactions at all scales and all levels of complexity was assumed by Newton and caused him to remark: "Nature is exceedingly simple and conformable to herself".[24]

---

[21] For a discussion of the genesis of Newton's insight see Westfall (1994).

[22] Strictly, it is the *net* external force acting on a body that causes a change in its motion. These topics will be further discussed in parts II and III.

[23] The intrinsic nature of forces, and even their very existence as physical entities, remain a subject of debate in some areas of philosophy and physics. See, for example, Bigelow et al. (1988) and Roberts (2015).

[24] Quoted in the introduction to the *Principia Mathematica* (Newton et al., 1999).

The vast body of physical theory that has evolved from this crucial insight, from classical to relativistic to quantum mechanics, considers such interactions only from an external, or extrinsic, perspective. The experiments on material systems that scientists conduct when studying mechanics are based on observations and measurements made from the outside in. Little, if any, thought has been given to what a material system feels from its intrinsic perspective during the interactions being studied or how any putative feelings it has might affect experimental outcomes. Rather, it has simply been assumed that it feels nothing.[25]

One person who *does* seem to have given thought to the inherent or intrinsic nature of matter during interactions between forces and bodies was Isaac Newton himself. In a much-discussed passage from the *Principia Mathematica*, definition III, he considers the inherent state of a body (a portion of matter) when it is acted on by an external force. This is such an important passage, and so concisely expressed, that it deserves to be quoted at length:

> A body, from the inactivity of matter, is not without difficulty put out of its state of rest or motion. Upon which account, this *vis insita* [inherent force], may, by a most significant name, be called *vis inertiæ*, or force of inactivity. But a body exerts this force only, when another force, impressed upon it, endeavours to change its condition; and the exercise of this force may be considered both as resistance and impetus; it is resistance, in so far as the body, for maintaining its present state, withstands the force impressed; it is impetus, in so far

---

[25] This attitude may be changing. For example, in his book about the Krebs cycle, the chemical process that drives metabolism, biochemist Nick Lane (2022) wonders whether the collapse of the membrane potential when a bacterium dies "'feels' like something to a bacterium". For related discussions see Reber (2018) and Ball (2023). Later in this treatise, the question of how the feelings of material systems might affect experimental outcomes will be discussed.

as the body, by not easily giving way to the impressed force of another, endeavours to change the state of that other.[26]

What Newton describes here is an antagonism or tension that occurs within the body between the resistance it offers to being changed, on the one hand, and its impetus to change the body that impresses the force, on the other. By using words such as "withstand", "difficulty", "persevere", "strive" and "endeavour", which appear elsewhere in his works and in different translations of this passage, Newton is alluding to a kind of will or volition that inheres in all 'active matter'.[27] Had subsequent physicists chosen to take this allusion of Newton's more seriously, the history of science—and indeed of thought more generally—might have been quite different.[28]

Interestingly, Newton concludes definition III with another example of 'dual-aspectism' by noting that while motion and rest are commonly considered to be opposing states they are in fact, as he puts it, "distinguished from each other only by point of view"—that is, the same object can be *both* moving *and* at rest, according to Newton, depending on the perspective that is adopted. This shows that the point of view or reference frame that we adopt with respect to a system when studying it is of fundamental importance in physics, just as it is in the form of psychophysics introduced in part II.6.

---

[26] Newton et al. (1999), translated from Newton's initial publication of 1687.

[27] In contemporary physics there is a growing field of 'active matter' research which studies how complex material systems behave when animated by forces and energy.

[28] In *Panpsychism in the West*, David Skrbina (2007) reminded us that Isaac Newton was not a strict Newtonian in that he did not regard the universe as a soulless machine driven by impersonal forces, in the way that Newtonians are sometimes said to do. Rather, he was led by his devout Christianity to regard the universe as imbued with soul and animated with spirit; like Gustav Fechner, he was a hylozoist. This aspect of Newton's thought will be discussed further in part III.2.

6. *What is it like to be a ball?* Newton's foundational insight that a body (portion of matter) must be acted on by an external (net) force if its motion is to change, and that this can be measured and mathematically modelled, can also be expressed in terms of work and energy transfer. In the case of the ball being kicked, kinetic energy is transferred from the boot to the ball as the boot does work on the ball. The question of what the ball feels when kicked will be addressed shortly.

I.8 Force, energy and work

Due to figures such as Gottfried Leibniz and Émilie du Châtelet (discussed in part III.4) and researchers such as James Prescott Joule, Hermann von Helmholtz and William Thomson (Lord Kelvin) who followed them, the eighteenth and nineteenth centuries saw scientists begin to describe many physical processes—particularly those involving mechanical systems and engines—less in terms of forces acting on bodies and more in terms of energy transferred and work done. The modern concept of energy, usually defined as the capacity to do work, has its origins in the thought of Aristotle but only emerged as a precise physical quantity during the eighteenth century.[29] By the nineteenth century it had become closely allied to the concept of work, formally defined as the application of a force to displace a system—that is, to change its motion—through a certain distance in the direction of the force by overcoming the system's resistance to change.

To express Newton's central insight about force and motion in terms of energy and work we can say that for an object to accelerate or change its motion it must be subjected to work in a process in which energy is transferred into or out of the object. Thinking of the ball mentioned earlier, the force applied through the foot when it is kicked does work on the ball, which acquires a certain amount of kinetic energy—the energy of motion—from the foot. The effect of the work done is to both accelerate the ball and to give it more energy.

The concepts of energy and work were originally developed to analyse the efficiency of mechanical systems such as steam engines, which convert sources of heat into useful motional work. It was soon realised, however, that the same 'thermodynamic' principles could be

---

[29] More strictly, the energy available to do work is termed the 'free energy', in contrast to the 'bound energy' of a system, which is not available for work. We will return to this distinction and to the nature of energy more generally in parts II and III.

applied to the study of more general physical processes like gases under pressure and chemical reactions, and even to biological systems, including systems like us. Eventually it was understood that all physical, chemical and biological processes, in one way or another, require a source of energy to perform work. Nineteenth century practitioners of neurophysiology, psychology and physics, including figures like Gustav Fechner, Hermann von Helmholtz and Ernst Mach, came to regard the human body, and indeed the brain and the mind, in quasi-mechanical terms as systems that absorb and dissipate electrochemical energy to perform useful biological activity.[30]

This energy-work-driven way of thinking about the operation of the world in general—and the body, brain and mind in particular—has been somewhat eclipsed in recent decades by the now-dominant computational paradigm that tends to treat biological systems like brains as information processors rather than energy–work systems.[31] But part of the purpose of this treatise is to demonstrate that the energy-work-driven approach that was dominant in the nineteenth century still has explanatory potential when integrated with modern information theoretical tools as we try to tackle the central problem of mind–matter relations today.

---

[30] One of the most prominent exponents of this tradition, besides Gustav Fechner, was Sigmund Freud. His uncompleted early work, *Project for a Scientific Psychology* (1895/1950), was an attempt to model the human mind and body as a thermodynamic system driven by tensions and forces and constrained by an economy of energy.

[31] This tendency is due in large part to the success of the digital computer and its impact on how we model complex systems and behaviour. See as a recent example of this information–driven approach the work of Li et al. (2022), which "offers a framework for modelling of brain dynamics in terms of information processing capacity". Meanwhile, one of the most prominent neuroscientific theories of consciousness is known as integrated *information* theory (Tononi et al., 2016). The value of energy-driven models of biological processes and their role in consciousness and vision are discussed in Pepperell (2018, 2020).

I.9   The experience of nature

What does it feel like to the ball when it is kicked, from its own point of view? What, if anything, does a portion of matter experience when it is subjected to work and energy is transferred to it? As noted above, Isaac Newton thought that it undergoes a tension or antagonism between two opposing tendencies: it resists or "withstands" being changed and "endeavours", in Newton's terminology, to oppose the external force by acting upon it with equal force to change it. This tension is the "inherent" condition of the matter that is acted upon; as philosophers might say, this is "what it is like" to be the matter from its intrinsic perspective.[32]

Even before Newton expressed the idea that matter experiences the effects of forces that act on it, figures such as Descartes had discussed interactions between bodies in quasi-psychological terms. And ever since the nineteenth century, physicists have routinely referred to the way objects and particles 'experience' or 'feel' the fields or forces they interact with. Chemists today often talk of the way certain reactions are 'favoured' or how a substance 'likes' a certain reactive outcome. They do not mean to imply that matter literally has psychological states; I have asked several and they say they do not. Yet this psychology-tinged language clearly offers an intuitive and convenient way to talk about the effect that forces have on matter, and as we will see, it may point to some basic fact about the nature of reality.[33]

---

[32] The philosopher Thomas Nagel (1974) once famously asked, "What is it like to be a bat?" He was referring to the fact that to be conscious is to be in a condition that feels like something for the agent concerned. The phrase "what it is like" is now widely used to define the subjective experience of a sentient agent.

[33] The physicist Ernst Mach saw no distinction between the mental and the material. In his book *The Science of Mechanics* he tells us, "We are ourselves a fragment of mechanics, and this fact profoundly modifies our mental life" (Mach, 1883/1960). Mach's ideas will be discussed further in part III.

Following Newton, it is here proposed that when a portion of matter is acted on by an external force, it experiences—in parallel to the action of that force—a tension from its own perspective. Put in energy–work terms: *when energy is transferred into a portion of matter during the performance of work, a tension is produced in the matter.*

Consider the case of the kicked ball using the psychophysical framework discussed in part I.6: the result of being kicked is that the ball has transferred to it a certain amount of energy due to work done, which is observed extrinsically as a physical event. At the same time, in parallel and unobserved, it is put into an antagonism, or what I will provisionally define in part II as a 'Conflict', which is felt intrinsically as a psychological event. From the perspective of the ball, this Conflict consists of a tension or stress due to its desire to maintain its original condition and the fact that it is compelled to change this condition by the Will of another system. Conflict and Will—which are also defined in part II—are proposed here as specific features of the general psychological property of matter that I will define as Experience. This Experience is transferred to the ball—intrinsically—in parallel to the energy which is transferred to it extrinsically.

There are, of course, other possible ways that psychological states might be explained in physical systems. Panpsychists claim that all particles of matter possess some basic level of consciousness from which more complex kinds, like our own, are built when these particles are combined in systems like brains. However, they have so far been unable to explain precisely what consciousness is, where or how it occurs in nature, how it can be measured, or how distinct conscious elements *combine* to form a unified mind.[34] It may be that consciousness exists

---

[34] This 'combination problem' has been described by, among others, William James (1890). Many contemporary panpsychists hold that *consciousness* is a fundamental and ubiquitous property of nature that is 'built in' to matter at the microscopic scale. For a recent anthology of representative views, see Goff and Moran (2022). The view

independently of matter and energy, as others have proposed, and that our brains 'tap into' or 'tune into' this source of 'cosmic consciousness' in some way we do not yet understand and which proponents of this view have yet to articulate. Others have claimed that consciousness comes into being because of the "objective" collapse of the wave function at a quantum scale of reality.[35] These and other suggestions, including mine, will remain conjectural until they are tested and repeatedly validated through experimentation.

To summarise the central idea being proposed here: the transfer of energy to a body or portion of matter—a process that can be measured extrinsically by observing its change in motion—is conjectured to entail the transfer of a general psychological property that exists only from the intrinsic perspective of the matter, which I am here calling Experience, specific features of which include Conflict and Will. Causally, the physical change in motion that is observed in the system occurs due to the familiar agents of change such as forces, charges and fields. The change in the psychological condition of the system occurs due to parallel causal agents that will be provisionally defined in part II but are yet to be experimentally studied. Nevertheless, these properties may be necessary to explain the behaviour of the system in full, especially if that system is a sentient agent.

So far, this tells us little about the nature of this property of Experience (which we will come to later), except that it entails a psychological tension for the matter and that being in this condition is what it is like for the matter to have that Experience.

---

being outlined here differs from this kind of panpsychism in that it treats consciousness as a special process of material systems that occurs only when a property defined as Experience is organised and behaves in sufficiently complex ways. This property is produced in matter under specific conditions and can be quantified and experimentally studied, as will be explained below.

[35] The most prominent exponents of this approach are Stuart Hameroff and Roger Penrose (2014).

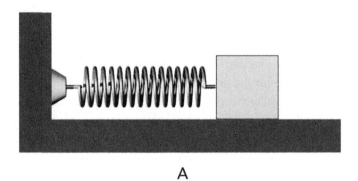

A

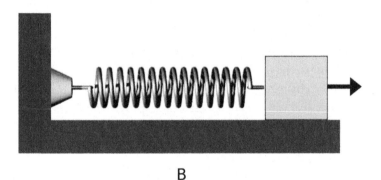

B

7. *What is it like to be a spring?* The spring–mass system shown in diagram **A** is relaxed and at equilibrium; no net forces are being applied to it, no work is being done and it has no elastic energy. When a force is applied to the spring as the mass is moved (shown in diagram **B**), work is done so that the spring is displaced from equilibrium and put under tension. During the work, energy is transferred to the spring that it stores as elastic potential energy.

I.10   The nature of Experience

To make the present suggestion more concrete and begin to show how it can be expressed quantitatively and used to generate testable hypotheses, consider an ideal spring, as in the spring–mass system illustrated in figure 7. When the spring is relaxed and no work is being done on it, the spring is at equilibrium. All forces acting on it are equalised and its motion relative to the world does not change. Extrinsically, no energy is being transferred to the spring. Intrinsically, it is without Conflict or Experience. As the spring is stretched, work is done to move it against the resistance put up by its stiffness. Kinetic energy is transferred to the spring during this action and stored in the form of elastic energy. This displaces the spring from equilibrium and so, intrinsically, it undergoes Conflict between its forced displacement and its wish to maintain its equilibrated state.

What does it feel like for the spring intrinsically when we observe it being stretched extrinsically? We might try to speculate or extrapolate from our own experience of being stretched, which often makes us feel uncomfortable. But because we have no direct access to the internal feelings of external objects when we observe them from the outside (including other people and animals), and because a spring is such a different kind of physical system to ourselves, we are limited in what we can infer about its experiences.

Even so, by applying the framework of psychophysical parallelism outlined in part I.6 and using the quantities and qualities defined later in part II, we can begin to infer a surprising amount about the feelings that occur in the spring in parallel to it being stretched and returning to equilibrium when relaxed. For example, when work is done on the spring (considered extrinsically), it is disturbed from equilibrium against its Will (considered intrinsically). This implies that it does not *want* to be put into Conflict and feels *displeased* or *uncomfortable* in being so changed. In psychological terminology, its Experience has *negative* hedonic valence. Conversely, as the work ceases (considered extrinsically) and the spring recovers its equilibrium in accordance with its Will, this implies that it feels *pleased* or *comfortable* to do so (considered intrinsically). Here, its Experience has *positive* hedonic valence.

In addition to what we can infer about the valence of the spring's Experience we can infer something about how strongly or intensely that valence is felt. So, the greater the work done and the more energy is transferred to the spring (considered extrinsically) the more *Intense* its feeling of displeasure (considered intrinsically) and the more Intense its pleasure at returning to equilibrium when relaxed.

While such inferences about the psychological properties of material systems are conjectural at this stage, most of them are not new— as we will see in the next section. And as previously stated, all these inferred properties will later be formally defined and methods of quantifying them proposed so that relevant hypotheses and predictions can be generated and experimentally tested, i.e., they are *scientific* proposals.

[As an aside, since we are considering an *ideal* spring here (for the sake of argument), we can ignore factors such as friction or heat and assume that we are dealing with an isolated system. This means that as the spring is stretched and relaxed, no energy is lost by the system to its surroundings. The process is fully reversible because the amount of stretching that is done—and hence the amount of displeasure experienced by the system—is equal to or *symmetrical with* the amount of pleasure it experiences as it is relaxed. This symmetry is shown in diagram **B** in figure 8 (i.e., the 'rising phase' is symmetrical to the 'falling phase') and will become important when we consider the implications of these ideas for living systems in part III.7. We will return later to the question of how relative amounts of pleasure and displeasure can be measured in material systems, even ones as simple as springs.]

I.11 The psychology of matter

The proposals made in the previous section will immediately meet with scepticism, if not outright rejection, in the minds of many people because I am attributing psychological states and agency to non-living objects like springs. But as noted, I am not the first to do so.

Hylozoism is the ancient and widespread belief that all matter has some form of agency, mind or life. Although it might now be largely rejected as mere superstition, many significant European scientists and philosophers in the modern period seriously advocated forms of hylozoism, including the idea that objects which are moved against their will feel displeasure and feel pleasure when they return to equilibrium. For example, the astrophysicist Johann Zöllner (echoing the quotes from Jeremy Bentham and John Locke in part I.1) said:

> All work done by organic or inorganic natural entities is determined by the sensations of pleasure and displeasure in such a way that the movements within a closed realm of phenomena behave as though they pursued the unconscious purpose of reducing the sum of unpleasant sensations to a minimum.[36]

In a similar vein, the biologist Carl von Nägeli wrote:

> Now if the molecules possess anything which is ever so distantly related to sensation, and we cannot doubt it, since each one feels the presence, the certain condition, the peculiar forces of the other, and, accordingly, has the inclination to move and, under circumstances, really begins to move, becomes alive as it were, moreover, since such molecules are the elements which cause pleasure and pain; if therefore the molecules feel something which is related to sensation, then this must be pleasure if they can respond to attraction and repulsion, i.e., follow their inclination

---

[36] Zöllner (1872), quoted in Arnheim (1971).

or disinclination; it must be displeasure if they are forced to execute some opposite movement, and it must be neither pleasure nor displeasure if they remain at rest.[37]

As noted in the previous section, such claims about the 'psychology of matter' remain conjectural and inferential at this stage. Part of the purpose of this treatise is to formalise such claims (like those made about valence and intensity), define the psychological properties in line with the form of psychophysical parallelism discussed in part I.6 and provide the conceptual tools necessary to design relevant experiments to test them. I stress again that the purpose of these formal definitions and tools is to generate falsifiable scientific hypotheses and predictions that can be experimentally studied in material systems such as nervous systems. By doing so, we may begin to explain how and why certain kinds of organisms have vivid and varied psychological states when we observe the matter and energy of their nervous systems behaving in certain ways.

In part III of this treatise, I present several case studies which show how these definitions and tools could be applied to help explain some long-standing problems in philosophy and science. Until these experiments are conducted and the underlying principles are experimentally validated and shown to be explanatorily useful, I acknowledge that what is proposed here remains conjectural.[38]

---

[37] Nägeli (1877). Gustav Fechner had previously expressed the same basic idea in terms of a "pleasure principle" that was later influential on the work of Sigmund Freud. See Heidelberger (2004). As will be discussed in part III.8, similar principles were proposed by founders of Gestalt psychology in the early 1900s.

[38] Stuart Hameroff (2017) has also proposed a physically based "pleasure principle" to explain the origins of life and the emergence of consciousness based on quantum mechanics rather than, as here, classical mechanics. See also the work of people such as Michael Levin (2019, 2021), who are considering what the minimal requirement is for a biological system to have agency.

I.12 Neuron–spring analogy

At this point the reader may be asking, "How do the supposed feelings of a spring or ball, even granting that they exist, have any bearing on the feelings I have in my nervous system when I am pleased or displeased?" The essence of the proposed answer can be introduced with the following analogy.

If we accept the 'neuron doctrine' that neuroscientists have traditionally subscribed to, which is that the brain functions due to the activity of individual neurons, then we can think of a single neuron as akin to a spring, albeit in a highly simplified way.[39] Neurons in their 'resting phase' are somewhat like a spring that is experiencing pressure or tension when it is compressed or stretched. There is a difference of voltage, or membrane potential, between the inside and outside of the neuron's cell membrane of around −70 millivolts which is maintained by a flow of charged particles of matter, or ions, across the membrane. These ions are moved against resistance, in part, by the biochemical work of microscopic pumps and channels embedded in the membrane that act to maintain the potential.

The effect of this membrane potential (due to the distribution of oppositely charged ions inside and outside the cell) is to put the body, or soma, of the neuron under pressure or tension because the ions want to balance their charges and return to equilibrium, but the largely impermeable membrane prevents them from doing so. The neuron in this condition is energetically 'poised' for action in much the same way

---

[39] The validity of the neuron doctrine, as conceived for a century or so, has recently come into question, not least because we are beginning to appreciate the functional importance of glial cells and intracellular structures such as microtubules, the role of connections between neurons and neuron ensembles, and because the belief that the only purpose of neurons is to act as binary signalling units is inaccurate (Guillery, 2005).

as a stretched spring held with a certain force is ready to contract but prevented from doing so by the holding force.

If the voltage potential across the membrane of the neuron decreases in magnitude to a certain level (around −55 millivolts) in response to the activity of neighbouring cells, then a dramatic sequence of events is triggered in the cell: its voltage potential shifts rapidly to zero during a rising phase, then becomes momentarily positive (peaking at around +40 millivolts), returns again to zero during a falling phase, drops to a negative value even lower than the starting potential (−90 millivolts or so), and finally returns to its original resting level of −70 millivolts, as shown in graph **A** in figure 8.

Driving this dramatic momentary fluctuation in voltage potential is a complicated set of reactions that carry out biochemical work on the matter of the neuron, the effect of which is to propagate a wave of electrochemical energy along the neuron's extended arm—or axon—that goes on to affect neighbouring cells. The neuron is able to actuate this energy transfer process—the 'action potential'—by harnessing the spring-like motive power of the opposing charges of its ions and their desire to move towards equilibrium whenever possible. This so-called firing process can happen many times a second in each neuron and is compounded in its effects on the nervous system as a whole by the collective action, or mass action, of thousands or millions of neurons firing in exquisite synchrony.

We can draw an instructive analogy between the action potential cycle in a neuron and the case of the stretched spring discussed earlier, albeit that the two systems start from different points relative to their equilibrium states. In its resting state, the spring has minimal energy and is not under tension; no work is being done to maintain it in this state. As the spring is stretched, however, it acquires energy during a rising phase, reaching a peak of tension when the maximum force is applied, then enters a falling phase as the stretching force abates and energy is divested, finally returning to its preferred state of rest, as shown in graph **B** in figure 8.

Despite the great differences between these two systems in terms of their material composition and organisational complexity, they nevertheless operate according to a shared principle, as is evident from the comparison illustrated in the figure below. Each system undergoes a cycle of 'energy–equilibrium fluctuation' in which it is driven away from equilibrium by the motive power of forces acting on, or within, the system when work is performed and then back towards equilibrium in accord with the intrinsic desire of the system to return to its preferred equilibrium state.

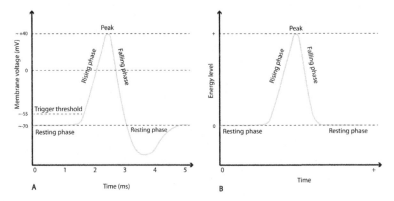

8. *The neuron–spring analogy.* Graph **A** shows the typical change in the voltage across the neuronal membrane during an action potential. An analogous process occurs in the case of the spring that is stretched and relaxed, acquiring and divesting energy in the process, as shown in graph **B**. Both processes involve resting, rising, peak and falling phases as they are moved away from and towards their respective states of equilibrium by forces acting on and within the systems when work is done. In each case, the system is undergoing a cycle in which it fluctuates between changes in its energy and equilibrium.

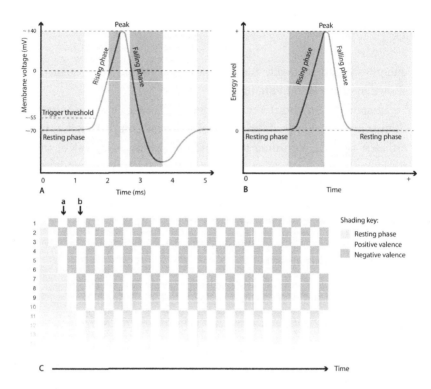

9. *Hedonic valence of material systems during cycles of energy–equilibrium fluctuation.* Graph **A** illustrates the different states of hedonic valence felt by a neuron during an action potential, as proposed here. Light shaded areas represent the stages of each cycle that have positive hedonic valence for the system as it tends towards equilibrium (zero voltage difference), while the dark shaded areas represent the sections or phases that have negative hedonic valence as the neuron is forced away from equilibrium. Mid shading represents the resting state. An equivalent but simpler expression of the same process occurs in the example of the stretched spring, as shown in graph **B**. Graph **C** shows a schematic ensemble of neuronal actions (1…*n*) that are synchronised so that at any point in time, such as that labelled **a**, the quantity of positively valenced phases exceeds the negative. The reverse applies for negatively valenced experiences. See part 1.14 for a discussion on the time point labelled **b**.

I.13   Feeling in nervous systems: valence

Thinking of a system's behaviour in terms of its energy–equilibrium fluctuation allows us to address the question of what it is like for an individual neuron to feel or experience an action potential from its own perspective. Referring again to the case of the spring discussed in part I.10, it was proposed that when forced from its preferred equilibrium state, the spring concurrently feels displeasure and that it feels pleasure when being allowed to return to that preferred state.

In psychology, these states of displeasure and pleasure are called negative and positive hedonic valence (as mentioned above) and are indicated by dark and light shading in graph **B** of figure 9. During the rising phase of the fluctuation, shaded in dark grey, the spring experiences negative hedonic valence as it acquires energy; as it divests energy during the falling phase, shaded in light grey, it acquires positive hedonic valence. Mid grey shading indicates the resting phase.

The analogous situation for the neuron, as illustrated in graph **A**, is more complex in operation but identical in principle. As the neuron moves from its resting to its rising phase—i.e., towards equilibrium—it experiences in parallel a positive hedonic valence which turns negative as it moves towards its peak—i.e., away from equilibrium—and becomes positive again at the start of the falling phase, then negative again, and briefly positive before settling at the resting phase, all these changes in hedonic valence being indicated by light and dark shaded sections. From this, we can infer the internal psychological condition of an individual neuron during each cycle of energy–equilibrium fluctuation by observing the phases of the action potential.

When considered at the level of a single neuron, these hedonic states may have little or no impact on the psychological condition of the nervous system to which it belongs, and hence the organism. But when states of hedonic valence are multiplied due to the synchronised interactions between millions of neural cells, the psychological effects on the wider nervous system may be significant. This hypothesis might help to explain how and why certain patterns of brain activity—as observed extrinsically—feel intrinsically pleasant or unpleasant to the nervous system, and hence its owner.

This hypothesis also leads to a prediction: the pattern of activity in an ensemble of neurons in the relevant region(s) of the brain of an organism that is experiencing pleasure, considered intrinsically, will be arranged such that at any moment, considered extrinsically, a greater quantity of light shaded than dark shaded sections occur simultaneously, as illustrated schematically in graph **C** of figure 9 (at time point **a**). In other words, there is a *net balance of positive valence* from the perspective of the neural matter concerned. The reverse applies for unpleasurable experiences: the dark shaded sections should synchronise in greater quantity at any given moment.

While it has long been known that the frequency or rate of neural firing is functionally important, it is now known that the precise timing—or 'temporal coding'—of firing is also critical.[40] With current scientific tools and methods, we should be able to test the above prediction in real and simulated nervous systems by observing, for example, the temporal coding of action potentials at the millisecond scale among populations of cells that are active in response to an unpleasant stimulus. In doing so, we would expect to observe synchronisation between phases of the fluctuations that are consistent with the negative valence evoked by the stimulus. Likewise, we should also be able to predict the intrinsic valence of the stimulus, and hence of the system's Experience, by extrinsically observing the temporal coding among the relevant neuronal populations. To my knowledge, such measurements of the hedonic valence of neural activity have not yet been conducted.

---

[40] For a recent overview see Andrade-Talavera et al. (2023).

I.14 Feeling in nervous systems: intensity

One of the most obvious things we learn from everyday experience is that in general, the more powerful the stimulus applied to our bodies the more intense the resulting experience. Referring to the kicked ball discussed earlier, if my leg were on the receiving end of the kick, I would normally feel a gentle kick less acutely than a firm one.

As with the kicked ball, the difference in stimulus intensity for a kicked leg can be expressed in terms of work done and energy transferred; in a gentle kick, less energy is transferred from the boot to the leg and so less work is done on the leg compared with a firm kick. This principle holds true—in general—for all forms of stimulation: we see brighter light when more photon energy enters our eyes to do work on our retinas; we hear a louder sound when more mechanical energy from acoustic waves enters our ears; we feel hotter surfaces when more thermal energy is transferred to cells in our skin.

I add the caveat "in general" because our sensory systems function such that we do not always perceive changes in the energy levels—or intensities—of stimuli. We are only sensitive to changes of intensity that occur within certain thresholds, and these vary with stimulus strength. For example, a person holding a 1 kilogram load to which is then added a further kilogram, thus doubling the load, will certainly perceive a difference. But if the person is holding a 25 kilogram load to which is added a further kilogram—which is a smaller fraction of the total—then the difference will most likely not be felt.

Gustav Fechner devoted much of his career to the study of these varying perceptual thresholds using the method of psychophysics mentioned earlier. This work was driven by his insight, which he verified experimentally, that there is a lawful relationship between the quantity of energy transferred to the sense organs and the perceived intensity of the stimulus.[41] Perceptual psychologists continue to study this relationship using the same psychophysical methods.

Investigations by physiologists in the 1920s revealed that the relationship between stimulus intensity and perceived intensity also relates to levels of energy transfer in neural tissue.[42] It was discovered, for example, that the rate at which sensory neurons in the peripheral nervous system fire their action potentials generally increases with the strength of the stimulus, a phenomenon now called 'frequency coding' or 'rate coding'. A sensory neuron responding to a firm kick on the leg will fire action potentials more frequently—thereby transferring more energy—than in response to a gentle kick.

Stimulation need not occur through sensory neurons for us to perceive sensations of varying intensity. Studies carried out since the 1960s have shown that by applying electrical current to the surface of the primary visual cortex—a region of the brain involved in processing vision—experimenters can evoke visual experiences of illuminated shapes, or 'phosphenes', in people who are totally blind.[43] The perceived number, size and brightness of the phosphenes can be increased by the application of stronger or longer currents, demonstrating that the quantity of electrical energy transferred to the cortex affects the intensity of the evoked visual experience.

Correlations between stimulus strength, perceived intensity and levels of energy transfer in neural tissue have been reported in all modes of sensation. In the domain of touch, for instance, researchers using electroencephalography (EEG)—a technique which measures electrical energy changes in the brain—found that increases in intensity of skin stimulation that were perceived more intensely produced

---

[41] Fechner (1860/1966). Fechner's insight, which came to him when waking on the morning of October 22nd, 1850, is celebrated annually by psychologists as Fechner Day.

[42] As reported in Adrian (1928).

[43] See, for example, Dobelle and Mladejovsky (1974).

quicker and larger voltage deflections in the sensory cortex of participants' brains that require more neural energy and work to produce.[44]

In the domain of hearing, researchers using fMRI—an imaging method that measures neural activity by tracking changes in energy consumption and distribution in the brain—found a positive relationship between the strength of a sound stimulus (measured by air pressure level), the subjective loudness of the sound as reported by participants, and the quantity of neural activation in the auditory cortex of the brain in terms of both the volume of tissue activated and the increase in levels of activity in that tissue.[45]

Results such as these might be explained, at least in part, by the form of psychophysical parallelism being outlined here. This suggests that in general, increases in the quantities of extrinsically observed energy flowing through a nervous system—and thus greater quantities of work being done on or in that system—parallel a greater intensity of Experience as felt by the system intrinsically.

Moreover, referring to graph **C** in figure 9, we might predict (as alluded to previously) that variations in the intensity of valenced experiences—i.e., the strength of their pleasantness or unpleasantness— may be a function of the number of light or dark shaded sections that occur simultaneously in the relevant areas of the nervous system. On this account, the intensity of the positively valenced Experience indicated at point **b** in graph **C** will be greater than that at point **a**, given the greater number of light shaded sections occurring in the system at once. As with the prediction made in part I.13, this could be tested empirically using real-time brain imaging methods such as 'voltage imaging', as illustrated in figure 10 and discussed in the next section.

---

[44] Mizukami et al. (2019); Park et al. (2021).
[45] Röhl and Uppenkamp (2012).

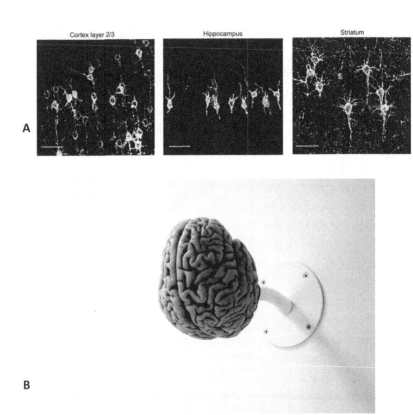

10. *Energy transfer and biophysical work in the nervous system.* Figure **A** shows sample images (by permission of Nature Research) from Piatkevich et al. (2019), wherein groups of active neurons from three different brain regions were illuminated by a voltage imaging technique. Figure **B** shows an artwork by the author called *Brain* (2008, resin and metal, 50 × 70 × 30 cm). The human brain contains around 86 billion neurons and a similar number of glial cells, most of which are electrochemically active at any one time in the way visualised in figure **A**. This activity is highly orchestrated by 'standing waves' that cause the brain to resonate with patterns of electromagnetic energy, much like the body of a musical instrument resonates when it is played, as discussed below.

I.15   Energy transfer in nervous systems

The neural energy transfer mechanisms discussed so far have been simplified to illustrate the proposed psychophysical approach. But when we apply that approach to the study of real nervous systems, we will encounter biophysical processes of far greater organisational and behavioural complexity. It is important to acknowledge this because, even though these processes *are* simple in the sense that they can all be described in terms of energy transfer in material systems, nervous systems have a vast repertoire of ways in which they can transfer energy to perform biophysical work (as noted in part I.2). Moreover, they can do so across an enormous range of scales, from the microscopic level within individual cells to the whole-brain level among large populations of cells.[46] We'll consider a few examples to illustrate the diversity of this repertoire and its breadth of scale; these same examples will be useful to bear in mind later when we come to discuss the processes that enable nervous systems to *consciously* Experience.

It is estimated that an adult human brain contains around 86 billion neurons and at least as many non-neuronal cells.[47] Each neuron is an extraordinarily intricate and exquisitely organised biochemical processing system, quite apart from the complexity of the work it might be doing in concert with thousands or millions of other cells. For example, each neuron is densely packed with elongated polymers called microtubules that perform several kinds of biophysical work, including exerting forces on the cell membrane to regulate its shape and supporting the transport of microscopic 'cargos' that are moved along neuronal microtubules by nanoscale motors.[48]

---

[46] This will be discussed further in part III.7 in relation to the work of Mae-Wan Ho.
[47] Herculano-Houzel (2012).
[48] Iwanski and Kapitein (2023).

Piatkevich and co-workers, as illustrated in image **A** in figure 10, used a method of optical voltage imaging in which modified genes are introduced into cells that cause them to exhibit variable fluorescence intensity depending on the membrane potential. This technique can track neuronal electrical activity such as that during an action potential, enabling researchers to observe the biophysical work performed by neurons in real time across many interacting cells.[49]

Meanwhile, at the whole-brain level there is growing evidence that highly orchestrated oscillations in neural activity coordinate behaviour across anatomically distant regions. Cabral and co-workers, for example, found that fluctuating patterns in the fMRI signals occurring in different parts of mammalian brains were strongly correlated, which they took as evidence for a 'resonance hypothesis' of brain function.[50] This hypothesis regards the brain as being closer to an acoustic musical instrument than a digital computer, in the sense that it resonates or vibrates when it is carrying out its biophysical work analogously to the way the body of an instrument does when it is played. These globally orchestrated patterns of neural oscillation are functionally important in cognitive tasks like memory and movement and are another of the many ways in which the brain transfers energy between its own matter when performing biophysical work.

When we observe a human brain from the outside, as represented by the artwork shown in image **B** in figure 10, we can see—depending on which imaging method we use—the many different and extraordinarily complex ways in which it transfers energy within itself to perform its work. What we do not see from that perspective, however, are the equally many and complex states of Experience that may occur in parallel to this activity from the brain's own perspective.

---

[49] Piatkevich et al. (2019). The power of the voltage imaging method is developing rapidly; for example, see Gilad (2024).
[50] Cabral et al. (2023). See also Atasoy et al. (2016) on 'connectome harmonics'.

I.16   Experience is not consciousness

So far in this treatise, a general framework and some accompanying principles have been outlined that might help us understand how certain feelings or Experiences occur in nervous systems in parallel with certain kinds of observed neural activity, and how they might be studied experimentally using current neuroscientific tools. But this leaves the central problem with which we started largely unresolved; namely, how and why they are *consciously* felt or experienced.

Whether due to sleep, anaesthesia or severe damage, the brains of people who are unconscious can still be highly active. In fact, in certain cases, neural activity as measured by metabolism (i.e., in terms of energy consumption) can be equivalent or even greater in the brains of unconscious than conscious people, including during sleep.[51] In all living brains, whatever the level of sentience of their owners, neurons are still undergoing action potential cycles, and large-scale, highly distributed areas of brain tissue are being subjected to biophysical work through the action of resonating oscillations. According to the principles outlined here, all of this should feel like something from the neural matter's own perspective. So why does this activity in the brains of people who unconscious not feel like something to their owners?

Before attempting to answer this question, it is necessary to say more about the nature of Experience and, in the succeeding section, about the differences in organisation and behaviour of neural activity between the brains of people who are conscious and those who are not.

---

[51] Dinuzzo and Nedergaard (2017); Bazzigaluppi et al. (2017). It is tempting to think of the non-conscious forms of Experience that occur in all living brains, as proposed here, in relation to various theories of 'unconscious experience' and 'unconscious drives', like those developed by Sigmund Freud and his followers (Freud, 1900/1913). See also the more recent work of Mark Solms in this context (Solms, 2021).

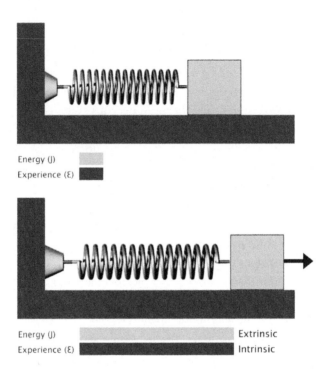

Energy (J)
Experience (Ɛ)

Energy (J)   Extrinsic
Experience (Ɛ)   Intrinsic

11. *The quantitative relationship between energy and Experience.* This diagram shows the same spring–mass system illustrated in figure 7 in both relaxed and stretched positions. As discussed below, is proposed that the amount of energy transferred to a portion of matter—in this case the spring—as observed extrinsically (light bar) is quantitively equivalent to the amount of Experience transferred to the matter from its intrinsic perspective (dark bar), as shown in this diagram. This proposal, I suggest, can be experimentally tested in living and non-living systems using units and methods that will be outlined in part II. Note that energy is measured here in joules (J), as is standard, while Experience is measured in a unit called Emps (symbol **Ɛ**), which will also be formally defined in part II.

I.17   The concept of Experience

The first point to clarify is, what exactly is meant here by 'Experience'? While we all know in one sense what the word 'experience' means, we need to define the term if we are to use it scientifically. Everyday words often have additional technical meanings and methods of measurement in science, energy and work being obvious examples. In part II of this treatise, I will propose a technical definition of Experience as a quantifiable property that can be used to design and conduct scientific experiments on living and non-living systems. Here, I introduce some key concepts that underpin that technical definition.

According to modern physics, a material system may have any number of quantifiable physical properties such as mass, volume, temperature, charge and energy. According to the framework being outlined here, a material system may also have several quantifiable psychological properties, the most general of which is Experience. Physical properties can be studied in systems when we observe them from an extrinsic perspective; psychological properties exist for systems from their own intrinsic perspective and cannot be directly observed. Nevertheless, both kinds of properties belong to the same system, as they do, for example, with us.

But how does this property of Experience relate to our more familiar conscious kind of experience? To answer this, we need to start by drawing an important distinction between 'Experience' and 'conscious Experience'. Experience—it is proposed—is a general psychological property of matter that occurs throughout nature whenever energy is transferred into or out of matter. In this general sense, it is a *non-conscious* phenomenon. *Conscious* Experience, as we shall see, is a particular type of Experience that occurs in certain material systems, such as the nervous systems of people and animals, when their matter and energy are organised and behaving in certain ways.[52]

The idea that Experience is a general property of matter may seem strange at first. But it is only a more formal and literal way of expressing what physicists routinely say when they refer to the 'experience' that a particle of matter has when a force or field acts upon it and changes its motion.

Returning to the example of the spring, when work is done to stretch it through application of a force, the spring is displaced and energy is transferred. Physicists might say that the spring 'feels' or 'experiences' the effect of the work being done, albeit without taking the psychological implications of these terms seriously. Here, I do take those implications seriously and say that when the spring acquires a certain amount of energy, which can be measured in any way energy transfer is normally measured, it *also* acquires in parallel a certain amount of Experience, as illustrated in figure 11. How can the quantity of this unobservable property be measured?

It is here proposed that we measure the amount of Experience a system acquires by giving it *quantity equivalence* with the amount of energy transferred to the system. This means that if the spring acquires 1 unit of energy as measured by observing its outward behaviour, it also acquires 1 unit of Experience from its intrinsic perspective. Put more formally: 1 unit of Experience is the amount of Experience

---

[52] The philosopher Alfred North Whitehead (1929) is one of several people to have drawn a distinction between experience as a general property of nature and conscious experience as a special form of that property. This view is sometimes called 'panexperientialism' and has some affinity with the concept of "mind-stuff" as proposed by William Kingdon Clifford (1878) and discussed by William James (1890) (see note 8). Note also that the psychophysical framework outlined here has been anticipated to some extent by William Seager's (2022) proposals on 'mental chemistry', John Stuart Mill's ideas about 'psychological chemistry' as discussed later in part II, and David Chalmers' (2007) proposals on 'double–aspect' principles.

transferred to matter when 1 joule of energy is transferred through work—the joule being the unit of energy in the International System of Units (SI), the worldwide international standard for units of scientific measurement. The rationale for the suggested name for the unit of Experience will be set out in part II, along with other key technical information about the proposed psychological properties of matter.

Admittedly, the proposed quantity equivalence between energy (in joules) and Experience (in Emps) is only conjecture at this stage, but it makes for a convenient starting point to guide future experimentation on the psychological properties of matter. We might also start with the assumption that there is a direct linear relationship between the amount of Experience acquired by a material system of a given mass and the Intensity of the Experience felt by that system, as suggested in part I.14, but the actual ratio may turn out to be different, perhaps taking the form of an inverse square law as is observed with light intensity or gravitational field strength over distance.

Even if it is shown by experiment that the equivalence proposed here is incorrect, the idea that Experience can be a measurable property of matter could, I think, prove to be a crucial step in advancing our scientific understanding of the mind. This is because, as noted in part I.6, it would put subjective mental states on a par with objective physical states by making them amenable to experimental study in material systems of all kinds, not only living systems. And given what was said in part I.13 about hedonic valence in the nervous system, it may point to ways of explaining the causes of pain and suffering in humans and other creatures based on their observable neural activity.

Most importantly for present purposes, it may open the way to a better scientific understanding—and indeed explanation—of conscious Experience itself. To make further progress in tackling this most profound of problems, we will first consider some pivotal recent knowledge about the neurobiology of consciousness.

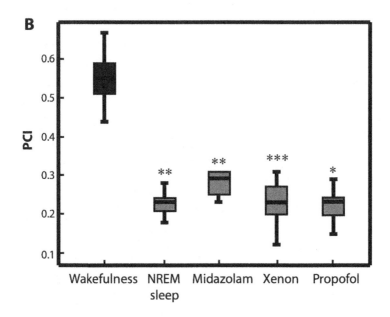

12. *Perturbational complexity index (PCI).* As discussed in the text, this key figure from a 2013 paper by Casali and co-workers shows the magnitude of the PCI in the condition where people were wakefully conscious (labelled 'Wakefulness') compared with when they were anaesthetised by one of several agents or when asleep. The authors claimed that this index, which is based on measuring the complexity of neural activity patterns in response to energetic perturbation, provides a way to distinguish between conscious and unconscious people based on observation of brain states. Courtesy of American Association for the Advancement of Science, 2013.

I.18   Conscious Experience and complexity

In 2013, Adenauer Casali and co-workers published a seminal paper that introduced an objective measure of a person's level of consciousness based on the pattern of their brain activity.[53] This measure, called the perturbational complexity index or PCI, was generated by first firing a pulse of energy into a person's brain using a technique called transcranial magnetic stimulation and then capturing an image of the brain's response to the stimulus using EEG, which measures the magnitude and distribution of the brain's electrical activity.

By comparing how mathematically complex the patterns in the EEG recordings were, which was done by measuring how far the data could be algorithmically compressed (akin to the way digital files are compressed to save memory), the researchers were able to show that brain responses exhibiting low levels of PCI, or complexity, were more likely to belong to people who were unconscious, whereas high levels were observed in the brains of conscious people (as shown in figure 12). In everyday language, more complex brain activity correlated with more consciousness. The PCI technique has been likened to hitting a bell and measuring the complexity of its vibrations to reveal its physical structure.

One reason this scientific result, which has since been replicated many times, was so important is that it allowed clinicians treating people with severely impaired brain function—many of whom are outwardly unresponsive to sensory stimulation—to better estimate whether they are minimally conscious. Another important study by Adrian Owen and colleagues had previously shown that some people with severely impaired brain function who were thought by doctors

---

[53] Casali et al. (2013). See also Massimini et al. (2022) for a more recent review.

to be in a vegetative condition could still understand and mentally respond to verbal instructions.[54] Clearly, it is a matter of the greatest ethical and clinical importance that such states of residual consciousness can be reliably detected.

Discriminating between people who are conscious and unconscious based on observable patterns of brain activity is now a well-established technique. Studies using different measures and methods have shown that the correlation between consciousness and complexity of brain activity is robust. Summarising this emerging "consilience of evidence" in a review paper in 2021,[55] Simone Sarasso and coauthors said:

> Overall, a large body of work supports the notion that the presence of consciousness is invariably associated with high brain complexity, which, vice versa, is found to be consistently decreased during physiological, pharmacological, or pathological-induced loss of consciousness.

In their review, these authors sampled 182 neuroscientific studies published mostly between 2010 and 2021 that had examined the relationship between consciousness and brain activity complexity. They found that even though the methods and tools used to conduct the studies varied greatly they shared a "common conceptual denominator", which was that they all tended to measure "the joint presence of functional integration and functional differentiation in the brain". I will explain what is meant by functional integration and differentiation shortly, but we will first review the history of how they emerged as explanatory concepts for conscious experience.

Key theoretical work in the 1990s by neuroscientists Giulio Tononi and Gerald Edelman introduced the concept that consciousness

---

[54] Owen et al. (2006).
[55] Sarasso et al. (2021).

is linked to the co-presence of differentiation and integration in brain organisation and behaviour.[56] Interestingly, this work was inspired by introspection on the contents of conscious experience itself. They observed that each specific experience—such as a moment of visual space perception—contains distinct parts (different patches of colour, texture, contrasts of brightness, edges, and so on) that are experienced as a unified conscious scene, as illustrated by the painting shown in figure 13. Taking this cue from the phenomenology of consciousness, they asked what kind of underlying anatomical and neurophysiological architectures might sustain it.

Their theory proposed three principles a system such as a brain must satisfy if it is to sustain conscious experience: (1) it is structured in a way that supports both differentiated and integrated activity; (2) it contains *re-entrant* anatomical processes that allow *loops of interaction* to occur between different subsets of the system in order to support their integration; and (3) it has a *repertoire* of possible differentiated neural states that can be integrated into specific experiences, with consciousness only being possible in systems having large, and therefore complex, repertoires.[57]

This theoretical framework turned out to be remarkably prescient in light of subsequent neurobiological evidence on consciousness and brain complexity. The review by Sarasso and co-authors showed that as the tools and methods of neuroscience advanced over time, researchers were able to ever more clearly reveal the extent of the brain's capacity for highly localised and widely distributed processing (i.e., differentiation), alongside its capacity for combining those distributed

---

[56] Tononi and Edelman (1998).
[57] The meaning of 'repertoire' has been developed in Tononi's subsequent work, but it is used here in the original sense that he and Edelman used it—i.e., the number of possible differentiated neural states that can be integrated into a conscious experience, large repertoires being synonymous with high complexity.

processes into larger functional units that sometimes span the entire brain (i.e., integration) with great variety and flexibility.

The evidence that accumulated through these studies supports the theory that it is only by achieving a finely tuned balance between large amounts of differentiation and integration that the brain sustains conscious experiences. Conversely, people whose brains fail to exhibit this balance have been consistently shown to lack consciousness.

Use of the term 'repertoire', which occurs frequently in this context, suggests a musical analogy for the kind of complexity being discussed. In music, repertoire refers to the entire body of works that a musician or musical group has the capacity to perform. A musician who can perform a great number of works of high stylistic diversity and historical range, for example, has a larger repertoire than one who can perform only a small number of stylistically similar works from a single historical period. Put in terms of complexity, the first musician's repertoire is more complex than the second.

In the studies analysed by Sarasso and coauthors, the complexity of the brain (its repertoire size) was measured by three distinct analytical approaches, mostly applied to data from brain imaging tools we have already encountered (fMRI and EEG): (1) detecting the number of discrete functional nodes or active regions operating in different *locations* in the brain ("topographical differentiation"); (2) the number of distinct processes occurring over *time* ("temporal differentiation"); and (3) combining these measures to detect how patterns of brain activity were distributed both *spatially and temporally*.

Having measured the diversity and distribution of various forms of neural activity, all the researchers computed the extent of functional integration using a range of mathematical tools. By comparing the resulting estimates of repertoire size, or complexity, in the brains of people having different levels of awareness, the link between complexity and consciousness gradually emerged from this large and methodologically diverse body of research.

It is worth pausing to reflect on what these measures of complexity in the brain mean for our intuitive understanding of the relationship between neural activity and consciousness. To extend the musical analogy, imagine a group of musicians who have identical instruments that generate very limited sounds, where the musicians know very few tunes, and where they are unable to hear or see each other and so cannot communicate or play in unison as they *ad lib* without following any common score. Our experience of hearing the music they perform, although it would be differentiated by the individual notes being played, would most likely lack the melody, harmony and rhythm we expect from a musical performance. It would be minimally integrated.

Now imagine the same musicians equipped with a wide range of instruments capable of expressing a variety of tones and timbres on which many notes can be played over many octaves, and where they know a great number of tunes, are highly aware of and attuned to each other's playing, and have rehearsed together extensively to follow any number of expertly composed and orchestrated scores. Our experience of hearing the music the group could now perform would likely be more differentiated and, especially, far more integrated, hence making it more melodic, harmonious and rhythmic than in the first case.

From this analogy we can appreciate the difference between a system having a small repertoire of states and one having a large repertoire that balances high levels of differentiation and integration. The musical richness we experience in the latter piece is analogous to the complex (highly differentiated) yet coherent (highly integrated) organisation of the brain in its consciousness-sustaining mode—at least according to currently available neurobiological evidence.

13. *An experience of visual perception.* This painting is titled *Self-portrait after Mach* (2012, oil on formed canvas, 100 × 150 cm, original in colour) and depicts the full scope of my visual experience when looking at my own feet with my left eye. Note that this visual experience is highly differentiated in that it contains many colours, shapes, textures, contrasts and edges yet fully integrated into a unique, single, coherent conscious experience. This painting is inspired by a drawing made by the physicist Ernst Mach in the 1870s showing a very similar view and illustrates what Giulio Tononi and Gerald Edelman in their 1998 paper (cited above) took to be axiomatic about the nature of conscious experience as revealed by introspection.[58]

---

[58] See Mach (1897/1914) and Clausberg (2007). For further discussion of the painting above and its implications for our understanding of visual perception and consciousness, see Pepperell (2024a). Part III.6 also includes a discussion of Mach's drawing.

I.19   From non-conscious to conscious Experience

In the context of the framework proposed here, the association of consciousness with complex brain activity may be explained with reference to the distinction made earlier, between (non-conscious) Experience as a general property of matter and conscious Experience as a special property of systems organised and behaving in certain ways.

We know that the nervous systems of animals, and their brains in particular, perform the complex biophysical work needed to execute tasks such as decision-making or coordinating bodily movement. This work is performed in multiple ways and at multiple scales from the microscopic level, involving microtubules and other subcellular structures within individual neurons, to the global level through long-range connections and widespread resonating oscillations. The energetic cost of performing all this biophysical work is notoriously high; the adult human brain consumes some 20% of the energy budget of the resting body while accounting for only 2% of its total mass.[59]

In the present framework, it is conjectured for experimental purposes that the performance of work on matter entails a parallel Experience from the intrinsic perspective of the matter. The energy demands of the brain show that it is doing a great deal of biophysical work and so must be producing large quantities of Experience of varying valence and intensity in its organic matter, compared with the rest of the body. However, we also know that the presence of this activity does not in itself mean that the owner of the brain is conscious of themselves or their environment; as we have seen from the neurobiological evidence reviewed in part I.18, this occurs only when the activity is sufficiently complex—i.e., it has a sufficiently large repertoire of differentiated and integrated states.

---

[59] Magistretti and Allaman (2013).

To reuse the phrase given prominence by Thomas Nagel, imagine "what it is like" for portions of organic matter in a human brain to carry out localised pockets of spasmodic, monotonous and disintegrated biophysical work; in other words, if the activity in the system expresses a small and narrow repertoire of organisational and behavioural states like the first piece of music discussed in part I.18. According to the form of psychophysical parallelism adopted here, the parallel states of Experience in the organic matter of that brain would be similarly localised, spasmodic, monotonous and disintegrated from the brain's point of view. In this scenario, there would be something it is like to be those localised active pockets of the system *individually and instantaneously*—due to the Experience each one has—but nothing it is like to be the system *collectively and continuously*. To refer again to the earlier musical analogy, the system produces *sounds* but lacks a *tune*.

Now consider a case where the brain activity has a very large repertoire of differentiated and integrated states, analogous to the second piece of music discussed previously. The many localised pockets of Experience occurring in different brain regions at different times are unified insofar as they are functionally integrated without erasing their local identity; they are organised into a recognisable tune. The brain has become highly complex (scientists might say it has passed a 'critical point' or undergone a 'phase transition', as illustrated in figure 16) in a way that is qualitatively different from the disintegrated state.[60] Its Experience, which was previously localised, spasmodic, monotonous and incoherent, is now unified, rhythmic, varied, orchestrated and temporally extended like the second piece of music described above.

---

[60] On phase transitions in physical and biological systems see Gitterman and Halpern (2004) and Heffern et al. (2021). A recent study has shown that the complexity of cortical electrodynamics, using the same measure as in Casali et al. (2013), enters a critical state or phase transition during consciousness (Toker et al., 2022).

During this greater complexity, where diverse regions of neural activity are harmonically resonating, a global 'melody of mind' is composed by integrating scattered 'sounds' of Experience. There is now something it is like, from their own point of view, for the relevant regions of the brain to experience diverse states synchronously and persistently; the system has entered the diverse and unified *conscious* phase of Experience as illustrated schematically in figures 15 and 16.

[As an aside, it is worth mentioning that the model of consciousness being sketched here is broadly compatible with some prominent neurobiological theories of consciousness, even though these established theories do not consider the intrinsic experiential properties of matter in the way set out in this treatise.

For example, the integrated information theory (IIT) of consciousness developed by Giulio Tononi, Christof Koch and co-workers[61] builds on the earlier work of Tononi and Edelman cited above. It proposes that for a system such as a brain to be conscious, its "information"—in this context meaning the specific way the system is differentiated—must be "integrated", meaning that it must be structured such as to form a whole that cannot be reduced to anything other than itself. In other words, every differentiated part in the system must be causally affecting every other part such that the system's "cause–effect repertoire", to use the authors' words, is maximal. Only systems reaching a certain threshold of irreducibility—which is measured mathematically using a theoretical quantity called Φ (phi)—can be conscious, according to IIT proponents.

A prominent alternative to IIT is the global neuronal workspace theory (GWT) developed by Bernard Baars, Stanislas Dehaene and

---

[61] Tononi et al. (2016). IIT has been highly influential in guiding the kind of empirical work reviewed by Sarasso and coauthors and also discussed above.

others.[62] According GWT's advocates, conscious states are produced when widely distributed unconscious neural processes—such as those concerned with memory, motor control, cognition, attention and sensory perception—become rapidly integrated into a "global workspace" in which all these distributed processes are made available, or "broadcast", to the entire system at once. Neurobiologically, this is achieved through "nonlinear ignitions" of neural activity that propagate through higher cortical areas in the brain. This 'ignited' activity is then amplified and sustained by *re-entrant* or *recurrent* neural processes of the kind referred to in part I.18 (and discussed further in part I.20) such that the information they carry can be globally broadcast among the widely distributed local processors. The claims of IIT and GWT are well supported by neurobiological evidence, and the theories are mutually compatible insofar as they claim the onset of consciousness depends on sufficient integration of otherwise disconnected neural processes. In IIT, this means crossing the threshold of Φ; in the case of GWT, it is the advent of global broadcasting following ignition.

Now consider the patterns of neural activity associated with consciousness in both theories, but from the intrinsic perspective of the nervous system itself and particularly its psychological property of Experience. The mechanisms of consciousness proposed by IIT and GWT, which are modelled on extrinsic observations of neurobiological processes, may help explain how the unobservable states that constitute the intrinsic psychology of the nervous system might transition from non-conscious to conscious Experience when sufficiently differentiated and integrated (as shown in figure 15). This explanatory approach could be studied experimentally by combining the tools and methods used by IIT and GWT with the technical definitions and concepts that will be outlined in part II of this treatise.]

---

[62] Baars et al. (2013); Mashour et al. (2020).

I.20   From conscious to self-conscious Experience

As noted already, one of the most important mechanisms for integrating differentiated and distributed neural activity is known as re-entrant or recurrent processing. Here, anatomically and functionally distinct regions of the brain share pathways, some extending over long distances, that send bidirectional signals (patterns of energy transfer) to generate neural feedback loops between regions.[63] These loops consist of populations of excitatory and inhibitory neurons that both stimulate/inhibit and are stimulated/inhibited by other populations. These recursive feedback mechanisms have long been implicated in the genesis of conscious processing and their operation is known to be impaired by anaesthetics.[64] Previously, I have likened the complexity that emerges from this self-referential looping to the extraordinary patterns that 'blossom' in video feedback systems when a camera views its own output.[65]

Imagine the complexity of the physical processes occurring in a nervous system with a differentiation–integration repertoire of incalculable size, operating at scales from the microscopic to the global, as recursive patterns of biological activity blossom. According to the form of psychophysical parallelism proposed here, these physical processes do not *cause* or *give rise* to the conscious Experience of the system. Rather, it is the parallel *psychological* processes occurring within the matter of the system, due to the inherent psychological properties of

---

[63] See Edelman and Gally (2013). These authors distinguish "re-entrant" neural processing from "feedback" as defined by mechanical engineers, which is as an error correction mechanism. I here use the term 'feedback' in its more general sense to refer to the recursive behaviour of a system that takes its own output as an input.

[64] For example, see Hudetz and Mashour (2016).

[65] See Pepperell (2018). The 'blossoming' that can occur in video feedback systems is another example of a phase transition. See the image in figure 21.

that matter, that constitute the conscious Experience from the intrinsic perspective of the system. The things we see, the pains and pleasures we feel are intrinsically experienced aspects of biophysical work occurring in the organic matter of our nervous system. Moreover, the fact that we experience these things *self-consciously* (being aware of our own awareness) may at least in part be due to the aforementioned self-referential and recursive organisation of our brains' exceptionally large repertoires of differentiated and integrated neural activity.

To better appreciate this, imagine that part X of a brain is experiencing certain 'something it is like' states and is sharing them with part Y, such that part Y experiences 'what it is like' to undergo the activity of part X as well as its own activity. This combined Experience in part Y is then fed back into the Experience of part X, albeit in modified form, and then back into part Y and so on, producing a self-reflecting feedback loop that results in the phenomenon of *Experience that is being experienced.* Here, the brain becomes a self-experiencing system that gives rise to higher-order experiential states that reflect upon lower-order ones—perhaps using data from sensory organs—intricately looping them together to form a whole which transcends the sum of its parts (see figure 14).

A mechanism such as this may underlie the definition of consciousness as "a sense organ for the perception of psychical qualities" given by Sigmund Freud in his book, *The Interpretation of Dreams.*[66] By this he meant that consciousness is a means of experiencing what is perceived by the external senses about the state of the world and by the internal senses about the state of the body; consciousness, on this view, is a higher-order self-monitoring and self-reporting mechanism that crowns a hierarchy of lower-order perceptions.[67]

---

[66] Freud (1900/1913).

[67] Several such 'higher-order theories' of consciousness have been proposed, e.g., by Lau and Rosenthal (2011).

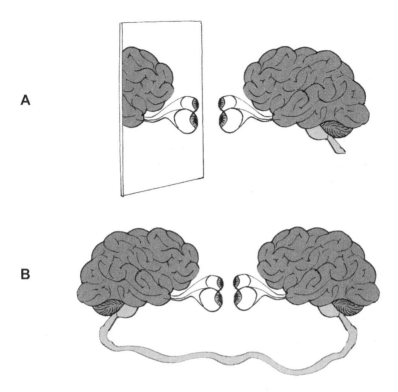

14. *Self-reflection and self-conscious Experience.* The illustration in figure **A** above represents a situation in which a psychophysical system that observes itself in a mirror experiences itself as a bundle of physical properties from an extrinsic perspective and, at the same time, as a bundle of psychological properties from an intrinsic perspective, as was mentioned in part I.5 and will be further discussed in part III.6. Figure **B** represents a psychophysical system that is observing its own experiential processes in such a way that its *Experience is being experienced,* as discussed above. It is proposed here that the recursive patterns of energy transfer that can arise from such self-observing and self-reflecting neural systems may 'blossom' into a higher-order organisation that is experienced, intrinsically, as *self-conscious* Experience.

15. *The principle of differentiation and integration in the transition from unconscious to conscious Experience.* In this highly simplified scale-free model, panel **A** shows a material system such as a nervous system, contained within the grey circle and consisting of several differentiated subsystems in which work is being done, indicated by the dots. According to the account given here, each of these subsystems is experiencing 'something it is like' states locally as it performs work. But as the subsystems are not spatially or temporally integrated, the repertoire of the system is far below the threshold sufficient for conscious Experience (CE), as indicated by the vertical bar on the meter in the lower left of the panel. Analogously to the first form of music described in parts I.18 and I.19, the system is insufficiently complex to be conscious. The graph in the bottom right of each panel plots as a dot the system's status in terms of the amount of differentiation (D, y-axis) and integration (I, x-axis) in the system, showing the threshold of sufficient repertoire for CE (dotted line). In panels **B**, **C** and **D**, the level of activity and the amount of differentiation and integration in the system gradually increase, indicated by the connecting lines as different layers of subsystems in different shades that increasingly interact. The repertoire threshold for CE is still not reached, but in panel **E**, additional interactions produce sufficient levels of activity and amounts of spatiotemporal differentiation and integration in the system to produce CE (indicated by the meter and the graph). Like the second piece of music discussed earlier being recognisable as rich and melodious, the system has a sufficiently complex repertoire to be conscious. In panel **F**, a distinct pattern of subsystem Experiences is integrated in a different way from panel **E** and represents a different CE for the overall system. Moreover, the greater contrast of the dots and lines that represent the subsystems and their interactions in panel **F**, and the more complex interactions between different layers of the subsystems, indicate that greater quantities of more complex work are being done; therefore, the CE is qualitatively richer and more intense than that in panel **E**.

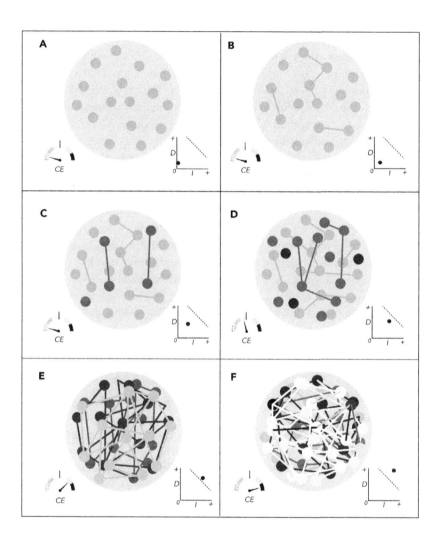

16. *Phase transition from unconscious to conscious Experience.* In this highly simplified diagram, the y-axis indicates the level of differentiation of the neural activity in a nervous system while the x-axis indicates the level of integration of its neural activity. As discussed in the text above, *differentiation* refers to the number of discrete parts a system has whereas *integration* refers to the way those parts interact with each other. Recent neurobiological experiments have established a clear association between the complexity of the differentiation *and* integration repertoire in brains and the presence of consciousness in the owners of those brains. We know that the onset of conscious Experience (CE) can be a gradual process, as, for example, is familiar when we wake slowly from a dreaming sleep; this is likely due to a gradual increase in the levels of differentiation and integration of neural activity. But we also know that it can be a sudden process, such as when we are rudely awakened from dreamless sleep. The diagram illustrates how, in principle, a system such as a brain may evolve from non-conscious Experience when levels of differentiation and integration are low—and neural activity is occurring without the presence of CE—towards a condition where CE is gradually dawning (referred to here as semi-conscious Experience), and further reach full or even heightened CE as the levels of differentiation and integration peak. The legend on the right notes some of the commonly used descriptions of these different states of conscious awareness. Note that in this model, the system can evolve from non-conscious Experience to CE in gradations but also by abrupt 'phase changes' that presumably occur with a more rapid increase in differentiation and integration to exceed certain critical values. Depending how fast the system evolves over time, the model proposed here can potentially account for both kinds of conscious-state emergence: i.e., slow—as opposed to rapid— evolution would tend to produce a more gradual emergence of CE.[68]

---

[68] Heffern et al. (2021), in their review paper on phase transitions, refer to first- and second-order transitions, the former being abrupt and the latter being more gradual. It seems probable that the emergence of consciousness in the brain can occur in both ways, or in combinations of both ways, depending on the circumstances.

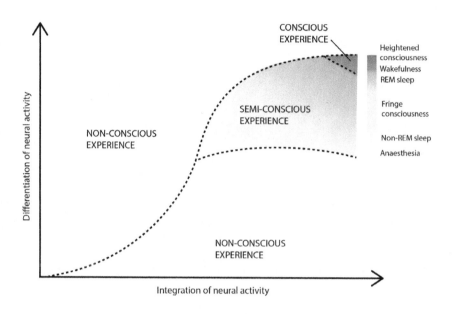

Coda 1

Part I of this treatise has outlined a conceptual framework for tackling one of the most difficult of scientific problems: experimentally studying and explaining the relationship between matter and mind. As stated in the preface, the depth and breadth of this problem demand the kind of pan-disciplinary approach taken here, even if the contribution from any single field can only be considered briefly. The benefit of this broad view, however, is that it has allowed us to draw tentative connections between fundamental principles of physics and the psychology of self-conscious Experience. Physics and psychology, which have so often been seen as remote—even isolated—poles of the scientific spectrum, become in this framework complementary ways of experimentally studying the same object or process, just as Fechner intended with his original conception of psychophysics.

It was also stated in the preface, and elsewhere, that this treatise does not aim to persuade by argument. Rather, it aims to provide context and rationale for the proposed framework, show that it has causal explanatory potential and a capacity to generate falsifiable hypotheses and testable predictions, and suggest ways of testing them. To that end, what follows in part II is a technical guide for those who wish to design and carry out such experiments.

I envisage two main ways in which these experiments could be conducted. One is to re-analyse or reinterpret existing datasets—whether they be from the 'hard sciences' (physics, biochemistry, neuroscience, etc.) or the 'softer sciences' of psychology, sociology, economics, and so on—by factoring in new psychophysical quantities and qualities such as those that will be defined in part II. The other way is to conduct experiments that are specifically designed to measure these newly defined quantities and qualities and analyse their role in the behaviour of the living and non-living systems being studied.

Part II

Technical guide

This part of the treatise summarises various definitions, principles, conjectures, quantities, methods of measurement and laws relating to the explanatory framework presented in part I. Besides providing more formal expressions of the concepts and properties introduced so far, part II defines further properties that may prove useful for those wishing to study the psychological states and behaviours of mechanical and biological systems. Importantly, the definitions and statements that follow are provisional and subject to experimental verification. They may therefore need to be revised. And nor is the list of definitions necessarily complete.

II.1   Primary conjecture

The primary conjecture on which this treatise rests is that *all matter can be considered to have intrinsic psychological properties*. This means that as well as the physical properties that we normally attribute to material systems—including mass, temperature and charge—we can assign various psychological properties, such as those provisionally defined here, the most general of which is *Experience*.

II.2   Definitions

Provided below are descriptions and definitions of concepts and properties from the proposed explanatory framework, along with suggested units and methods of measurement where appropriate.

Definition 1. *The system*

A *system* is any part of the universe—i.e., a portion of matter—being studied.

Definition 2. *Extrinsic perspective*

Any system can be considered from two parallel perspectives for the purposes of scientific study. One is the *extrinsic perspective*, which is its physical nature as observed from an external viewpoint with respect to the system. The properties of a system studied from the extrinsic perspective are generally regarded in science as being observer-independent. Strictly speaking, however, the extrinsic perspective presumes an observer and so is observer-dependent to some extent (see part I.6), although this can often be ignored for experimental purposes.

Definition 3. *Intrinsic perspective*

The *intrinsic perspective* of the system is the psychological nature of the system considered from its own viewpoint. This cannot be observed from an extrinsic perspective with respect to the system, so its existence must be inferred based on what can be observed extrinsically (making it also observer-dependent to some extent). The psychological properties defined herein exist only from the intrinsic perspective of the system and, following the form of psychophysical parallelism discussed earlier in part I.6, exist in parallel to and are identical with the physical properties that are observed extrinsically. The intrinsic and extrinsic perspectives are two ways of viewing the same thing.

Definition 4. *Parallelism*

Parallelism in this context means that certain psychological properties of material systems, which exist only from the intrinsic perspective of the system, complement or parallel certain physical properties that are measured from its extrinsic perspective using standard methods and units. From these measurements, the psychological condition of the system can be inferred. In some cases—as with *Nolition* (defined below)—the psychological property parallels more than one physical property, so the parallel can be approximate or its measure can be contextually dependent on the nature of the system being studied (e.g., whether a mass, spring or wave system).

Definition 5. *Experience*

Experience is a general (extensive and scalar) property of a system that refers to its intrinsic psychological condition when being disturbed from or recovering its preferred state of Repose (as defined below). Experience can be considered as the psychological parallel of what is measured extrinsically as the energy (E), or the change in energy (ΔE), of the system when work is performed on or in the system. In some cases (as discussed in part III.4) it may be useful to consider the interconversion of energy from one form to another, as measured extrinsically. Quantitatively, 1 unit of Experience is the amount of Experience possessed by, transferred to or interconverted within a system when 1 joule of energy is possessed by, transferred to or interconverted within the system during work. The unit of Experience is the Emp (short for the Greek for experience, *empeiria*) with the symbol $\mathbf{\mathcal{E}}$ (Latin capital letter epsilon, bold). Its quantity is given by the equation $\mathbf{\mathcal{E}} = E$, where E is the energy in joules as measured using any suitable method, or ΔE in cases where energy change is being measured. While Experience is a *general* property of material systems, there are several *specific* properties as will now be defined.

Definition 6. *Repose*

In classical mechanics, a system or part of a system is in mechanical equilibrium when all forces acting on it are balanced and no acceleration or perturbation is occurring in or to the system, i.e., no work is being done on the system. As observed extrinsically, the behaviour of a system will tend towards mechanical equilibrium when work ceases, which is often a continuously static state. In some situations, as in complex chemical or biological systems, there can be a stable, dynamic or steady equilibrium where forces continue to act in or on the system (and work continues) but there is a balance of energy flowing into, within and out of the system. This is often a continuously moving process. Considered from the intrinsic perspective, the desired or preferred condition of a system is *Repose*, which is the psychological parallel of equilibrium of whatever kind. Repose is symbolised by bold lower-case epsilon, ε, from the first letter of the Greek word for peace, *eirene*). In general, a system having minimum Experience has maximum Repose, given its circumstances.

Definition 7. *Will*

Will is a property of the system that refers to its desire to maintain or recover its preferred state of Repose, ε, and to affect any other system that attempts to disturb it from this state. In classical mechanics, Will can be regarded as the psychological parallel of the physical mass of a massive system, which is a quantifiable measure of its inertia, in kilograms. However, the parallel physical measure is context-dependent. In the case of stretching or compressing an elastic material such as a spring, Will is the parallel of the spring constant (measured in Newtons per metre) that quantifies its resistance to change in length. When considered in the context of neural activity, Will plays a role wherever biophysical work is being performed in the nervous system; that is, where portions of organic matter are displaced, reconfigured or reorganised during energy transfer processes. The unit of Will is the Thelon (from the Greek for will, *thelema*), with the symbol **T** (Greek capital letter tau, bold). Quantity equivalence is again proposed: when measuring mass, 1 **T** = 1 kg, or when measuring a spring constant, 1 **T** = 1 N/m.[69]

Definition 8. *Nolition*

Nolition is a measure of how much the system is disturbed from its preferred state of Repose contrary to its Will when interacting with another system. In classical mechanics, Nolition is the approximate psychological parallel of what is measured extrinsically as the acceleration of a system, where acceleration is the rate of change in velocity, but like Will, it is context-dependent. In the case of stretching or compressing a spring, it would be parallel to the displacement or change in the length of the spring; in the case of a wave in a medium, it would be parallel to the amplitude of the wave. In the context of neural activity, Nolition plays a role wherever the actions of forces or energy transfers on organic matter in nervous systems, or parts of those systems, are being analysed. Thus, the amount of Nolition that a system undergoes intrinsically is given by measuring the acceleration of the system, the extent of its displacement, or its amplitude. The unit of Nolition is the Nolon with the symbol **N** (Greek capital letter nu, bold), such that 1 **N** = 1 m/s^2 (in classical mechanics) or 1 **N** = 1 m (for a spring or wave).

---

[69] Will can also be considered as the psychological parallel of a force in the same way that Newton, in Definition III of the *Principia*, treats inertia as the *vis inertiæ* or "force of inactivity" of a body that acts to maintain its state of motion by resisting the force impressed by another body. This point becomes significant when considering the proposed Laws of Experience, as set out below. Interestingly, the contemporary physicist Matt Strassler (2024) defines mass as an object's "intransigence" to motional change; intransigence is normally defined as a psychological property.

Definition 9. *Conflict*

Conflict is a property of the system that refers to its psychological condition when it is disturbed from its preferred state of Repose contrary to its Will when interacting with another system. Conflict can be considered as the psychological 'tension' or 'stress' that exists in the system when it acquires Experience contrary to its Will.[70] In classical mechanics, Conflict can be regarded as the psychological parallel of what is measured extrinsically as the force experienced by the system. The unit of Conflict is the Agon (from the Greek for conflict, *agon*) with the symbol $\mathbf{X}$ (Greek capital letter chi, bold). It can be found by measuring the net force acting on the system, such that $1\ \mathbf{X} = 1$ kg m/s^2. Unlike force, which is a vector quantity with magnitude and direction, Conflict is a scalar quantity that has magnitude but no direction from the system's intrinsic perspective, making Conflict the parallel of the *magnitude* of the force acting on the system. Conflict is proportional to the product of Will and Nolition, such that $\mathbf{X} = \mathbf{T} \times \mathbf{N}$.

Definition 10. *Volition*

Volition is a property of the system that refers to the direction (either attraction or repulsion) in which it wishes to exercise its Will with respect to another system to achieve its preferred state of Repose. The symbol for Volition is $\mathbf{\Upsilon}$ (Greek capital letter upsilon, bold), and it can take a positive or negative value depending on whether the system attracts ($\mathbf{\Upsilon}^-$) or repels ($\mathbf{\Upsilon}^+$) another system, as observed extrinsically. Considered intrinsically, Volition represents the inherent desire of a system to achieve union with or separation from another system.

---

[70] Maxwell (1877) refers to the antagonism that exists between the forces of two interacting bodies as "Stress", which implies the psychological quality that such interactions entail for each body. But since Stress is now a technical term used elsewhere in physics, the term Conflict is used here instead to convey the same quality.

Definition 11. *Distress*

Distress is the negatively valenced hedonic quality that a system feels as it acquires Experience and is disturbed from its preferred state of Repose. Distress is the derivative of Experience with respect to the *direction* of the energy change where this has a positive value, irrespective of duration. The unit of Distress is the Algon (from the Greek for pain, *algea*) with the symbol **A** (Greek capital letter alpha, bold). Its value is given by the change in Experience, $d\boldsymbol{\mathcal{E}}/dt$ or $\dot{\boldsymbol{\mathcal{E}}}$, where $\dot{\boldsymbol{\mathcal{E}}}$ has a positive value ($\dot{\boldsymbol{\mathcal{E}}} > 0$).

Definition 12. *Relief*

Relief is the positively valenced hedonic quality that a system feels as it divests Experience and moves towards its preferred state of Repose. Relief is the derivative of Experience with respect to the *direction* of the energy change where this has a negative value, irrespective of duration. In some cases when free energy is divested, Relief can be considered as the approximate psychological parallel to the change in entropy of the system (see note 72). The unit of Relief is the Hedon (from the Greek for pleasure, *hedone*) with the symbol **H** (Greek capital letter eta, bold). Its value is given by the change in Experience, $d\boldsymbol{\mathcal{E}}/dt$ or $\dot{\boldsymbol{\mathcal{E}}}$, where $\dot{\boldsymbol{\mathcal{E}}}$ has a negative value ($\dot{\boldsymbol{\mathcal{E}}} < 0$).

Definition 13. *Intensity*

Intensity is the relative strength of a system's Experience proportional to the Will of the system, where Will is measured by a quantity such as the mass or the spring constant. Put in intrinsic qualitative terms, a system into which is transferred a given amount of Experience will feel that Experience with greater Intensity the smaller is its Will; put in extrinsic terms, a system on which a given amount of work is done will have a greater quantity of Intensity the smaller its mass. The unit of Intensity is the Kraton (from the Greek for strength, *kratos*) with the symbol **K** (Greek capital letter kappa, bold), and its value is given by the Experience divided by the Will: $\mathbf{K} = \mathbf{\mathcal{E}} / \mathbf{T}$.[71]

Definition 14. *Affliction*

Affliction is the Intensity of the Distress that a system suffers. Qualitatively speaking, this is how strong the negatively valenced feeling of Distress is for the system in proportion to its quantity of Will. Affliction has the symbol $\mathbf{K_A}$ (derived from the symbol for Intensity), and its value is given by the quotient of the system's Distress and Will: $\mathbf{K_A} = \mathbf{A} / \mathbf{T}$.

Definition 15. *Delight*

Delight is the Intensity of the Relief that a system enjoys. Qualitatively speaking, this is how strong the positively valenced feeling of Relief is for the system in proportion to its quantity of Will. Delight has the symbol $\mathbf{K_H}$ (derived from the symbol for Intensity), and its value is given by the quotient of the system's Relief and Will: $\mathbf{K_H} = \mathbf{H} / \mathbf{T}$.

Definition 16. *Shock*

Shock is the rate at which a system acquires Distress. Qualitatively speaking, a system that acquires a given quantity of Distress over time t will feel twice as much Shock than if that same quantity of Distress is acquired over time $2t$. Shock is the derivative of Distress with respect to time. It has the symbol $\dot{\mathbf{A}}$ (derived from the symbol for Distress) and is measured in units of Algons per second ($\mathbf{A}/\mathrm{s}$).

---

[71] The property of Intensity as defined here is distinct from the use of 'intensity' as equivalent to radiant energy flux in contemporary physics. Maxwell (1877) also refers to "intensity", by which he seems to mean the relative strength of a force. Note also the distinction between *Intensity*, which is proportional to Experience and Will, and *Conflict*, which is proportional to Will and Nolition.

Definition 17. *Joy*

Joy is the rate at which a system acquires Relief. Qualitatively speaking, a system that acquires a given quantity of Relief in time t will feel twice as much Joy than if that same quantity of Relief is acquired over time $2t$. Joy is the derivative of Relief with respect to time. It has the symbol $\dot{\mathbf{H}}$ (derived from the symbol for Relief) and is measured in units of Hedons per second ($\mathbf{H}/s$).

There are some general points to note about the properties defined above. First, following Newton's statement in Definition III as cited in part I.7 that the *vis inertiæ*, or force of inactivity, is only exerted when another force is impressed, the properties defined here—and Experience in general—are assumed to inhere in a system only when it is interacting with another (see the First Law of Experience, below).

Second, the Latin and Greek symbols used to denote the properties are displayed in capital bold type to indicate they refer to the *intrinsic psychological perspective* of the system, distinct from the conventional symbols used in physics that are displayed in regular type and refer to the *extrinsic physical perspective*. This convention may prove helpful when both perspectives are being discussed or studied at once. The quantities, states, properties, symbols and relationships defined here are summarised in table 1 and figure 17.

Finally, it is worth saying that this attempt to define psychological properties of nature was anticipated by the philosopher John Stuart Mill in the 1860s when he proposed that mental phenomena might be explained as products of "psychological chemistry", as discussed by William Seager (2016). We are a long way from achieving a "mental chemistry" level description of our minds, but we may be closer to explaining some basic feelings like pleasure and pain as products of 'psychological physics', or psychophysics, as proposed here.

Table 1.

Psycho-logical quantity	Parallel physical quantity	Unit	Sym-bol	Formulae
Experi-ence	Energy (or energy change, ΔE)	Emp	$\mathcal{E}$	$\Delta\mathcal{E} = \Delta E$ (general) $1\,\mathcal{E} = 1\,J$ $\mathcal{E} = \frac{1}{2}\cdot m\cdot v^2$ (kinetic) $\mathcal{E} = \frac{1}{2}\cdot k\cdot v^2$ (elastic potential) $\mathcal{E} = m\cdot g\cdot h$ (gravitational potential)
Repose	Static or dynamic equilibrium	Emp	ε	$\varepsilon = \Delta E = 0$ (static) $\varepsilon = dE/dt = 0$ where $\mathcal{E} > 0$ (dynamic)
Will	Mass (m) or spring constant (k)	Thelon	$\mathbf{T}$	$1\,\mathbf{T} = 1\,kg$ (mass) $1\,\mathbf{T} = 1\,N/m$ (spring)
Nolition	Acceleration (a) or displacement (x) or amplitude (A)	Nolon	$\mathbf{N}$	$1\,\mathbf{N} = m/s^2$ $1\,\mathbf{N} = m$ $1\,\mathbf{N} = m$
Conflict	Force (F)	Agon	$\mathbf{X}$	$\mathbf{X} = \mathbf{T} \times \mathbf{N}$ (general) $\mathbf{X} = F = m\cdot a$ (mass) $\mathbf{X} = F = k\cdot x$ (spring)
Volition	Attraction or repulsion (− or +)	Volon	Υ	Υ^- or Υ^+
Distress	Direction of energy change, if positive	Algon	$\mathbf{A}$	$\mathbf{A} = \dot{\mathcal{E}}$ if $\dot{\mathcal{E}} > 0$

Psycho-logical quantity	Parallel physical quantity	Unit	Sym-bol	Formulae
Relief	Direction of energy change, if negative	Hedon	**H**	$\mathbf{H} = \dot{\boldsymbol{\varepsilon}} \text{ if } \dot{\boldsymbol{\varepsilon}} < 0$
Intensity	Quotient of E and mass (m) or spring constant (k)	Kraton	**K**	$\mathbf{K} = \boldsymbol{\varepsilon} / \mathbf{T} \text{ (general)}$ $\mathbf{K} = \boldsymbol{\varepsilon} / m \text{ (mass)}$ $\mathbf{K} = \boldsymbol{\varepsilon} / k \text{ (spring)}$
Affliction	Quotient of positive energy change and mass (m) or spring constant (k)	Kraton	**K$_A$**	$\mathbf{K_A} = \mathbf{A} / \mathbf{T} \text{ (general)}$ $\mathbf{K_A} = \mathbf{A} / m \text{ (mass)}$ $\mathbf{K_A} = \mathbf{A} / k \text{ (spring)}$
Delight	Quotient of negative energy change and mass (m) or spring constant (k)	Kraton	**K$_H$**	$\mathbf{K_H} = \mathbf{H} / \mathbf{T} \text{ (general)}$ $\mathbf{K_H} = \mathbf{H} / m \text{ (mass)}$ $\mathbf{K_H} = \mathbf{H} / k \text{ (spring)}$
Shock	Rate of positive energy change in the system	Algon	**$\dot{\mathbf{A}}$**	$\dot{\mathbf{A}} = \mathbf{A}/\mathrm{s}$
Joy	Rate of negative energy change in the system	Hedon	**$\dot{\mathbf{H}}$**	$\dot{\mathbf{H}} = \mathbf{H}/\mathrm{s}$

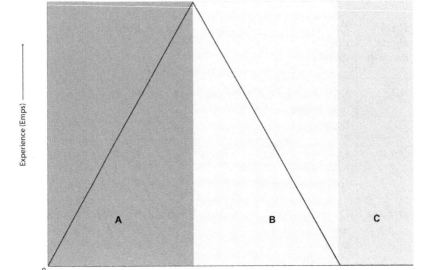

17. *Valence phases of Experience.* This graph shows a plot of imaginary data for a system which accumulates and divests Experience (y-axis) over time (x-axis), to illustrate the different qualitative valence phases of that Experience from the system's intrinsic viewpoint. The system starts at its preferred state of Repose where its quantity of Experience, measured in Emps, is zero. During the dark shaded phase, labelled **A**, the system undergoes Distress, measured in Algons, as it acquires Experience ($\dot{\mathcal{E}} > 0$). During the light shaded phase, **B**, it undergoes Relief, measured in Hedons, as it divests its Experience while returning to its preferred state of Repose ($\dot{\mathcal{E}} < 0$). During the mid shaded phase, **C**, it has minimal Experience having maximised its Repose (e). Note that phase **A** in this graph corresponds to the rising phases shown in the diagrams in figure 9 for the neuron and the spring, phase **B** to the falling phases, and phase **C** to the resting phases. The rate of change of Distress and Relief is measured as Shock (Algons per second) and Joy (Hedons per second).

82

II.3   Psychophysical conjectures and principles

The following conjectures and principles are offered as guidance for those conducting scientific investigations into the psychophysical behaviour of material systems within the framework outlined here.

The conjectures are further statements about the nature of material systems that follow from the primary conjecture of the framework: *that all matter can have psychological states.* These conjectures are intended to be helpful when considering the behaviour of material systems, forming research questions and hypotheses that aim to explain their behaviour, designing experiments to test and falsify hypotheses, and interpreting their results.

The principles may govern the psychophysical behaviour of material systems, both observable and inferred. They are intended to allow the experimenter to make predictions about the outcomes of their experiments and to explain behaviour in a principled way. These conjectures and principles are provisional statements and are subject to modification in light of experimental results.

Conjecture 1: For every physical change that is observed in a material system there exists a parallel, yet identical, psychological change in the system, the nature of which can be inferred from the observation.

Conjecture 2: For every observed physical change there is physical cause, and for every inferred psychological change there is a psychological cause, both of which are aspects of the same cause.

Conjecture 3: In every case where energy is transferred to or from a material system, as observed extrinsically, there is a parallel and proportional transfer of Experience from the system's intrinsic perspective, both of which are aspects of the same transfer.

Principle 1: A system, as observed physically, changes its motion to an extent that is proportional to the net force exerted on it and inversely proportional to its inertia; considered psychologically, it changes its Nolition to an extent that is proportional to the Conflict it suffers and inversely proportional to its Will.

Principle 2: Every system—as observed physically—tends, where possible, to minimise the energy it acquires and maximise the entropy it produces; considered psychologically, it endeavours to minimise its Distress and maximise its Relief.[72]

Principle 3: The direction in which a system changes when acted on by a net force, as observed physically, is in line with the direction of the net force; considered psychologically, a system made to suffer Conflict by another system endeavours to satisfy its Volition.

---

[72] The implications of this principle for the evolution and behaviour of living systems will be considered in part III.7. Note that there are at least three different forms of the quantity "entropy" that are discussed in contemporary science—thermodynamic, statistical and information theoretical—and authors are not always explicit about which form, or forms, they are referring to. The concept of entropy, including its different meanings, will also be discussed in more depth in part III.7. In this Principle, I use entropy in the thermodynamic sense as defined by Rudolf Clausius in the mid-nineteenth century when developing his mechanical theory of heat. Entropy from this viewpoint is a measure of the transformation that occurs in a heat-driven mechanical system when it loses some capacity to do useful work, that is, when it divests free energy. This transformation tends to be irreversible in nature, and so entropy always tends to increase. As Clausius (1867) succinctly put it: "(1) The energy of the universe is constant. (2) The entropy of the universe tends to a maximum". Finally, it should also be noted that David Chalmers (2007) suggests, as part of an outline theory of consciousness, a series of psychophysical principles that have some affinities with those presented here. Chalmers' theory, which also takes experience to be a fundamental property of nature, proposes that phenomenal properties reside in the intrinsic aspect of information processing rather than, as here, energy processing. However, information and energy are deeply interrelated in this context (see Pepperell, 2018).

II.4 Extrinsic and intrinsic laws

The following text and tables present the well-established *Laws of Motion* and *Laws of Thermodynamics* expressed in their conventional form, conceived from the extrinsic perspective of material systems. Alongside each law is presented a parallel version that is conceived from the intrinsic perspective of material systems. Together with the conjectures and principles just given, these statements aim to provide the experimenter with a fuller understanding of the nature of the psychophysical systems they are investigating and the causes of the behaviour they observe.

II.4.1 The Laws of Motion and Experience

Isaac Newton's Laws of Motion, first published in the *Principia* of 1687, express one of the most profound and far-reaching ideas ever conceived: that the observed motions of material systems can be precisely described and predicted using mathematical methods. These laws and methods have been of incalculable importance ever since their publication. As is the convention in science, however, they were formulated only with respect to the extrinsically observable motions of bodies. In table 2, I offer a provisional set of laws—the Laws of Experience—that parallel Newton's Laws of Motion by considering the intrinsic perspective of a body or system and the nature of its Experience during interactions.[73]

As noted above, the parallel between physical and psychological properties in the proposed framework can be approximate. Consequently, translating the extrinsic concepts expressed in the Laws of Motion into intrinsic Laws of Experience is not straightforward; indeed, it is necessary to think quite differently about what occurs *experientially* in a material system compared with what occurs *mechanically*. Motion, for example, plays no part in the Laws of Experience whereas it is the very essence of Newton's laws.[74] And although they are yet to be experimentally verified, the fact that the Laws of Experience, and the First Law in particular, can be expressed somewhat more succinctly than the parallel Laws of Motion suggests that the former may be more general than the latter.

---

[73] The intrinsic expressions of the Laws of Motion could also be called the 'Laws of Emotion' in reference to their qualitative and psychological nature. In this table I use the English translation of the Laws of Motion in the *Principia* as presented by Cohen and colleagues (in Newton et al., 1999).

[74] See the reference below to the physicist Ernst Mach's claim that the "sensitivity" of matter is more general than its "mobility", in part III.6.

Table 2.

Laws of Motion	Laws of Experience
First Law (Extrinsic)	**First Law (Intrinsic)**
Every body perseveres in its state of being at rest or of moving uniformly straight forward, except insofar as it is compelled to change its state by forces impressed.	Every body perseveres in its state of Repose, except insofar as it is compelled to suffer Conflict.
Second Law (Extrinsic)	**Second Law (Intrinsic)**
A change in motion is proportional to the motive force impressed and takes place along the straight line in which that force is impressed.	Nolition is proportional to the Conflict suffered by a body and inversely proportional to its Will.
Third Law (Extrinsic)	**Third Law (Intrinsic)**
To any action there is always an equal and opposite reaction; in other words, the actions of two bodies upon each other are always equal and always opposite in direction.	For any act of Will there is always an opposing act of Will; in other words, two bodies that interact will each suffer the same Conflict.

II.4.2 The Laws of Thermodynamics and Psychodynamics

The Laws of Thermodynamics can be expressed in several different ways. They govern the behaviour of material systems that are exchanging energy within themselves and with their surroundings or converting energy into different forms.[75] As with the Laws of Motion, existing statements of the Laws of Thermodynamics consider only the extrinsic perspective of the system. In table 3, I present the three most cited laws in their conventional extrinsic form and in a parallel form that considers the intrinsic perspective of the system. I have termed these (provisional) intrinsic laws the 'Laws of Psychodynamics' in reference to the field of psychology of the same name (to which Sigmund Freud was an early contributor) that studies the effects of forces on mental behaviour.

The Laws of Thermodynamics were originally developed to describe the behaviour of engines that convert heat to mechanical work by exploiting the motive power of temperature gradients. The concept of thermodynamic entropy was developed to describe the loss or dissipation of free energy available to do work in the system as it performs a cycle. But more recently it has been applied to the study of biological systems, and—of most relevance here—neural processes.[76]

The occurrence of Relief, as defined in part II.2, approximately parallels the increase in the entropy of a system. This is because it refers to the tendency of a system to minimise its free energy where possible—a process which is positively valenced for the system—to achieve its preferred state of Repose (i.e., static or dynamic equilibrium), depending on the nature of the system and the prevailing conditions. This will be discussed further in part III.7.

---

[75] The Second Law of Thermodynamics will be further discussed in part III.7.

[76] For an example relating to the neural basis of emotion see Déli et al. (2022).

Table 3.

Laws of Thermodynamics	Laws of Psychodynamics
First Law (Extrinsic)	**First Law (Intrinsic)**
The energy of an isolated system can be converted to different forms, but the total quantity of energy cannot change.	The Experience of an isolated system can be converted to different forms, but the total quantity cannot change.
Second Law (Extrinsic)	**Second Law (Intrinsic)**
In any interaction between a system and its surroundings the total quantity of entropy will never decrease. The entropy of the universe will always increase.	A system will never increase its quantity of Experience on its own; where possible it will always decrease its quantity of Experience.
Third Law (Extrinsic)	**Third Law (Intrinsic)**
The energy of a system at absolute zero will tend to an absolute minimum.	The Experience of a system approaching Repose tends to a minimum.

II.5 Experimental principles

Below are outlined some provisional principles by which, for experimental purposes, the psychophysical behaviours of material systems can be measured using the properties defined in this framework. These principles are intended to offer further guidance for those conducting experiments on systems having wide-ranging levels of complexity, both living and non-living.

II.5.1 Simple spring–mass systems

Consider again the spring–mass system discussed in part I. To measure the Experience of the spring when relaxed, being stretched and relaxed again, we simply need to convert the physical quantities we obtain when measuring the spring constant—the force applied (in N), the energy transferred as work is done to the spring (in J) and the displacement (in m)—into the corresponding psychological quantities and qualities summarised in table 1. From this conversion, as well as by considering the timing of the stretching and relaxing movements, we could calculate the change in Experience in the spring and its Conflict over time, quantify the valenced qualities of Distress and Relief it undergoes, and determine how Intense its Experience is. From these measurements made from the extrinsic perspective of the system as it changes its physical states, we can infer a good deal about the changes in its psychological states.

II.5.2 Multi-element spring–mass systems

Consider a more complex experimental setup in which several masses are connected to each other with springs to form a network. Perturbation of one mass in the network (by transferring energy to the mass) causes other masses in the system to be perturbed through displacement of the connecting springs. Although the calculations of the physical states of elements in the network—that is, the energy transfers in each mass and spring and the accompanying displacement or change in length of the latter—would be considerably more involved in this composite system than in the first case (and more so the more elements the system contained), the same approach of conversion from physical to psychological units would apply as in the first case. Thus, an equally detailed account of the changes in the system's psychological states as it is perturbed could in principle be obtained.

II.5.3   Complex material and biological systems

The example of the multi-element spring–mass system illustrates an important general principle concerning the study of the psychological states of composite systems, however many components they have and however complex the interactions of the components. To study the psychology of a system like that illustrated in figure 16—to measure, for example, whether it is experiencing greater Distress or Relief—we need to combine the measurements of energy transfer in each component or group of components. There are several ways this could be done, such as by summing the average of the quantities or the changes in these quantities of the various properties for each component (see the techniques discussed in parts I.13 to I.15). But whichever approach we take, we are effectively *integrating* the *differentiated* states of the network through measuring the combined effects of their energy transfers to and from each other and in the system as a whole—as discussed above in part I.18 with respect to measures of neural complexity. By applying this same principle to the analysis of ever more complex systems of energy transfer, including living systems and nervous systems, we can experimentally study ever more complex psychological states. Essentially, the advice to experimenters is: "Follow the energy and the Experience will follow!"

II.5.4   Some examples of potential experimental paradigms

As further guidance to potential experimenters, I will briefly mention some experimental situations—in addition to those discussed in parts I.13 and I.14 concerning the measurement of perceived hedonic valence and intensity—where the framework set out here might offer additional explanatory power.

Of particular interest in this context is the work being carried out in studying 'dissipative structures' in physics and biology.[77] Dissipative structures are systems that tend to configure themselves by self-organising to minimise the free energy that flows into them and maximise the entropy they produce. These systems are observed to break the symmetry between inward and outward flows of free energy (unlike the ideal spring discussed in part I) and so are irreversible thermodynamic processes. Fascinatingly, in doing so they often display life-like organisation and behaviour, even for relatively simple mechanical systems such as a collection of metal beads in oil that is subjected to a voltage change.[78]

Such phenomena relate to the theory of 'dissipative adaptation' developed by Jeremy England and colleagues, who have made similar observations in simulated systems of the kind mentioned in part II.5.2 above.[79] These findings are generally rationalised by reference to modern theories of nonequilibrium thermodynamics, which are still relatively new and not fully understood, and of course consider only the extrinsic perspective of the experimental systems. The framework outlined here has potential to be able to predict this surprising behaviour in nonequilibrium systems—and perhaps more fully explain it—by considering the psychological states of the systems concerned.

---

[77] Kondepudi and Prigogine (2015).
[78] Kondepudi et al. (2020). Note that the concept of free energy used here is distinct from, but related to, the information theoretical form of free energy (Friston, 2010).
[79] England (2015, 2020).

Coda 2

This outline of the proposed approach to one of the deepest of problems—as given in the treatise so far—is not intended to convince anyone of its viability or merits in and of itself. It is merely an invitation to consider the psychological states of material systems for experimental purposes, and to carry out experiments within the framework defined here with the aim of more fully explaining our observations of the behaviour of those systems. In particular, the aim is to causally explain psychological states by inference from observed physical states by making principled predictions, as outlined above, and falsifiable hypotheses.[80] I hope that I have done enough to persuade at least some people that the experiments are worth conducting and the predictions are worth testing in real and simulated systems.

In part III of this treatise—after a brief interlude of images and comments—I will provide more concrete illustrations of the rather abstract ideas discussed so far, using some well-known and some less well-known cases from the history of science and philosophy, and reflect on what they might mean for other deep problems such as the nature of life and aesthetics. In doing so, I hope to 'road test' the explanatory potential of this framework by throwing new light on these historical cases and deep problems.

---

[80] One way that the hypotheses proposed in this treatise could be falsified would be to show that conscious Experience can be produced in a system that is not subject to transfers of energy; in other words, in a static or inactive system. The neuroscientist Giulio Tononi and collaborators have argued that according to their integrated information theory (IIT) of consciousness, a 'silent' or inactive system of logic gates would be conscious if correctly arranged (Tononi et al., 2016; see also Bartlett, 2022, for a critique of this argument). The framework proposed here predicts that any conscious system requires the presence of energy-driven activity, and so it directly contradicts IIT in this respect.

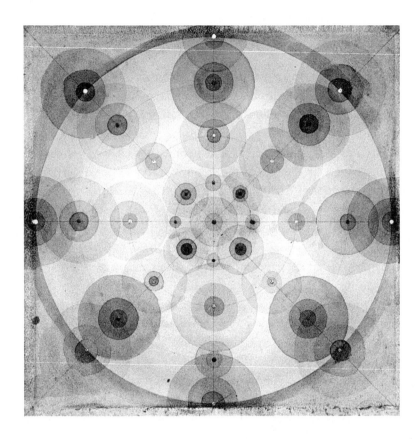

18. *Lines of force XIV*, 2021. Ink and pencil on watercolour paper, 40 × 40 cm (original in colour). This painting is part of a series inspired by Michael Faraday's experimental and artistic work on the invisible fields of electromagnetic energy that permeate the space around matter, which he called "lines of force". It illustrates in a highly simplified way how fields of force surrounding 'point centres' diminish in intensity as they spread out and interact with other fields to form complex geometrical structures. As discussed below in part III.5, some people think of brain activity, and consciousness itself, in terms of interacting fields of electromagnetic forces.

19. *Cymatic drawing (Chladni figure)*, 2023. In 1787, the musician and physicist Ernst Chladni published images of patterns formed in sand on metal plates that were vibrated using a violin bow. These Chladni figures, like the one shown here, are examples of the 'cymatic' patterns created when energy flows through matter, as explored by Hans Jenny and discussed later in part III.7. Given what has been outlined in this treatise, we can imagine the brain as a kind of 'organic plate' that is vibrated in highly complex ways by electromagnetic fields and other sources of energy; in which case our conscious experiences reflect the patterns formed in that plate, as experienced from our intrinsic perspectives.

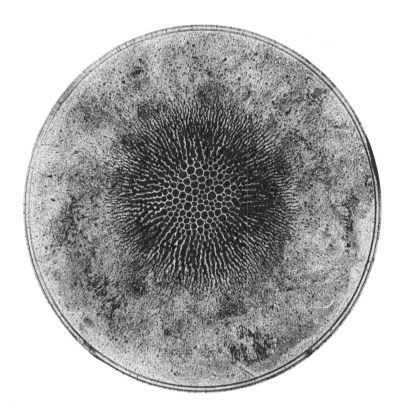

20. *Field lines*, 2020. Ferrofluid, ink and resin in petri dish, 6 × 6 cm (original in colour). This artwork was made by suspending ferrofluid in resin and allowing it to set when resting on the pole of a powerful magnet. The patterns are produced by interactions between the fields of energy surrounding the magnet and the iron particles in the fluid. It is an example of how seemingly organic patterns can emerge as particles of matter find stability when interacting with fields of force, as discussed in part III.

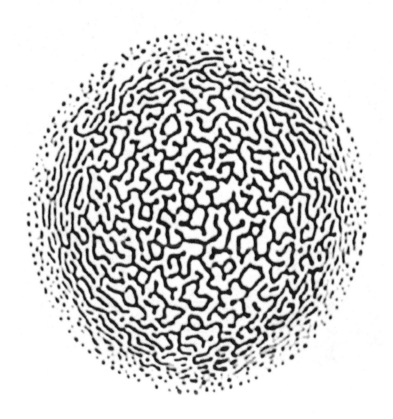

21. *Video feedback drawing X*, 2018. Computer drawing. This drawing is based on a still from a video feedback sequence created by the author. Video feedback can occur when a camera is pointed at a screen showing its own output. What starts as a tunnel-like effect can 'blossom' into ever-changing patterns of great complexity and beauty. The first time I encountered video feedback, as a VJ in the 1980s, my immediate thought was: "This is what consciousness must look like!"

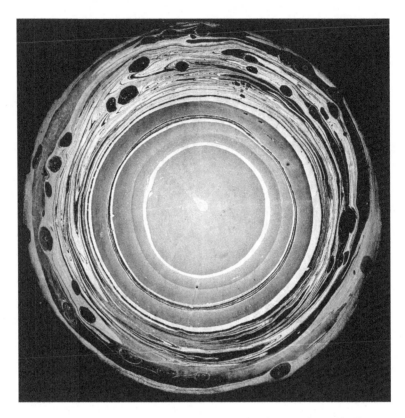

22. *Energy may be called the fundamental cause for all change in the world (Werner Heisenberg)*, 2024. Marbled paint and chalk on paper, 60 × 60 cm. Although his work is not discussed in this treatise, Werner Heisenberg (who is quoted in the title of this piece) was another scientist who, like Albert Einstein, Erwin Schrödinger, Niels Bohr, Wolfgang Pauli and Max Planck (all mentioned elsewhere in this treatise), contributed to the development of quantum physics in the twentieth century. In his book *Physics and Philosophy* (1958), Heisenberg equates the modern concept of energy in physics to the "fire" that the early Greek philosopher Heraclitus claimed was the cause of all activity in nature.

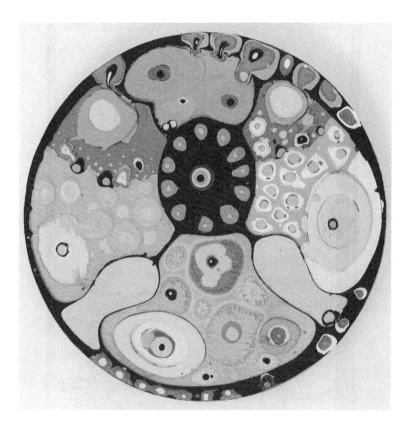

23. *Energy is the capacity of a system, acted upon by forces, to experience a specific amount of change (Eugene Hecht)*, 2021. Jesmonite on panel, 60 × 60 cm (original in colour). The title of this work quotes a statement made by a leading physics educator in a paper on the nature of energy. It is an unusual definition of energy, but one that succinctly captures the essence of the thesis outlined in this treatise, that Experience is an intrinsic property of material systems that parallels extrinsically observed changes or flows of free energy. All the matter in this artwork experienced change as it was acted on by forces while the work was being made.

24. A drawing from René Descartes' *Treatise on Man*, first published in 1662, which illustrates the physiological processes that occur when a person touches a flame. We assume that something causes a person to move their hand into or out of the hot flame, but is that cause mental or material in nature? Philosophers since at least the time of Descartes have debated this question.

Part III

III.1   Elisabeth of Bohemia's mental causes

René Descartes has gone down in philosophical history as the person who, more than any other, is associated with dualism: the idea that the mind and body are entirely separate and irreconcilable kinds of substance. The persuasiveness of his arguments in support of this view have been acknowledged ever since they were published in the seventeenth century. They remain pervasive to this day and in part are responsible for the challenge we now face in trying to integrate mind and body within one explanatory framework. But as noted in part I.4, Descartes' reputation on mind–body dualism is somewhat undeserved as he also explicitly asserted the *union* of mind and body.

The logical discomfort of Descartes' position on mind–body relations was exposed during a lengthy and famous correspondence with Elisabeth of Bohemia. She was a minor royal and a highly educated and well-connected intellectual figure living in the Netherlands and Germany who corresponded with Gottfried Leibniz, among many others, and in later life became an abbess. She read Descartes' *Meditations* (1641) shortly after it was published, met him at The Hague, and by 1643 they had begun to exchange letters.

In one early letter, Elisabeth sought clarification from Descartes on a central problem raised by his philosophical work. How is it, she asked, that the soul (or what we might now call the mind) can cause voluntary actions to occur in the body if, as Descartes argued, soul and body are two entirely different kinds of substance?[81] Souls are immaterial substances and bodies are material ones, according to Descartes, and while she could readily understand how a material object could cause motion in another material object by mutual contact, she could not conceive how the same effect could be caused by something immaterial, such as the soul, which makes no contact with the body.

In the philosophical literature, this is now known as the problem of mental causation and is in one sense the inverse of the problem, discussed in part I.6, of how physical events in the brain 'cause' mental states such as consciousness. As modelled in figure 25, it is often assumed that the intent to touch an object (shown in the dashed circle a') causes a physiological event (such as the brain's motor preparation response indicated in the solid circle a), which then causes the bodily action of touching the object (as shown in the solid circle b) with the consequent feeling of doing so (in the dashed circle b'). But this sequence, and the ones that follow in the remaining circles, entail the same sort of explanatory gaps between physical and psychological causes that were discussed in part I.6, again indicated in each case by x. Philosophers still debate this problem today.[82]

Descartes' response to Elisabeth was to say that while it had been necessary for didactic reasons to make an initial distinction between soul and body and to attribute to them only what properly pertains to each, he had not yet explained how soul and body are united and how one influences the other. This he then attempted to do using analogies, such as the way gravity acts on a material body without any apparent

[81] Descartes (1989).
[82] As examples see Crane (1995) or Robb et al. (2023).

physical contact. Elisabeth was not convinced, and despite their lengthy correspondence it seems the matter was never satisfactorily resolved.

The form of psychophysical parallelism that underpins the present framework suggests a solution to this problem—at least for experimental purposes—which is modelled in figure 26. As with figure 4, the sequence of events in question has parallel physical and psychological aspects that respectively have physical and psychological causes and effects.[83] Only the former can be observed empirically in each case; the latter must be inferred. Yet to fully explain the sequence of events we must consider them as a single causal chain.

Observing the physical aspect, we can attribute cause and effect to the physical behaviour of forces, masses and energy transfers. Considering the psychological aspect, we can attribute cause and effect to the 'psychodynamics' of psychological properties such as Conflict, Will and Experience, as defined in part II. So, for example, the mental 'pressure' exerted by the wish to touch an object is caused by the transfer of certain quantities of Experience to one part of the nervous system from another. This pressure is then relieved as the Experience is transferred elsewhere in the system in accord with the system's Will and Volition.

As with figure 4, this model lacks explanatory gaps with respect to causation. And as with figures 3 and 4, the scientist studying the process can only observe the physical aspects of this sequence of events from an extrinsic perspective. Note also that the causal chains indicated by the dashed and solid arrows in figure 26 are represented such that they can be considered separately and jointly. Ideally, they would

---

[83] The left–right orientation of figures 25 and 26 is mirrored compared with figures 3 and 4, for the convenience of allowing circle a' in figure 25 to appear first and make the causal order of the sequence clearer.

be shown diagrammatically in superposition, where they are simultaneously identical and complementary.

What would Elisabeth and Descartes have made of the model proposed here that attempts to answer the question they debated for so long? Perhaps it would have helped if they had thought of what they called the immaterial as what we might now call the 'unobservable', i.e., an aspect of nature that is just as real as the material, but which exists only from the intrinsic perspective of a system and is therefore unseen to us as observers. It may also have helped if they had applied a model of causality which is closer to that proposed by Leibniz in which material (i.e., observable) bodies are not moved through the causal agency of mental (i.e., immaterial/unobservable) ones, and nor are mental events caused by material ones. They may then have surmised that, as proposed in part I.6, material and mental causes are two aspects of the *same* causal agency, one producing observable (material) effects and the other producing unobservable (immaterial) effects.

It is also tempting to imagine how neuroscience would look today if parallelism of the kind proposed by Spinoza, Leibniz or Fechner rather than Cartesian dualism (which is effectively what is modelled in figures 3 and 25) had become the dominant philosophical framework for understanding brain-body and mind-matter relations. If so, would we now be closer to explaining the relationship between neural activity and consciousness? It is possible that we would. To his credit, Descartes proposed some ingenious mechanisms to explain our responses to external stimulation in the posthumously published *Treatise on Man*.[84] In discussing the physiological processes that he illustrated in the woodcut shown in figure 24, in which a person touches a hot flame, he talks of "spirits" that are released by the stimulation of sense receptors and which force open "tubes" that travel to the brain to

---

[84] René Descartes (1662/1972).

"press" on other tubes there—much as parts of a machine press on each other—resulting in bodily motion and associated "emotion" when "united" with the soul via the pineal gland.

With some generosity we can reinterpret Descartes' "spirits" in more modern terms as electrochemical energy that is transferred through "tubes", or nerves, to do biophysical work on other "tubes", or neurons, to bring about motions and emotions. His proposal about the pineal gland has long since been discredited. But what if he had he adopted some form of psychophysical parallelism instead of dualism, treating those "spirits" as the invisible but no less real 'mind side' of bodily actions? Perhaps the explanatory power of his mechanical model—and our contemporary understanding of psychophysics— would both be much greater.

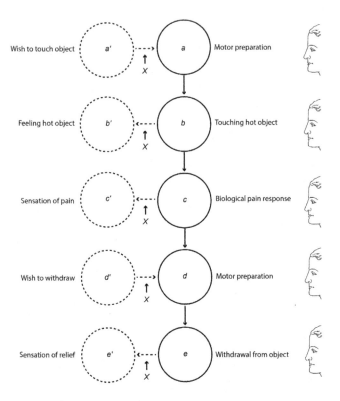

25. This figure shows a model of the commonly assumed relationship between psychological states (dashed lines) and physical events (solid lines) in science and philosophy. The psychological event of wishing to touch an object (circle *a'*) supposedly causes an observable motor preparation event in the brain (circle *a*) that causes the act of touching (circle *b*), in turn causing the sensation of feeling the hot object (circle *b'*). The sequence of events continues (circles *c* to *e and c'* to *e'*) with causal relations between psychological and physical events as indicated by the dashed and solid arrows. Explanatory gaps that exist at each step are indicated by **x** in every case. Note that in common with figure 3, this model is closer to the dualist view than those in figures 4 and 26, since it treats mind and matter as two distinct realms connected by unknown mechanisms.

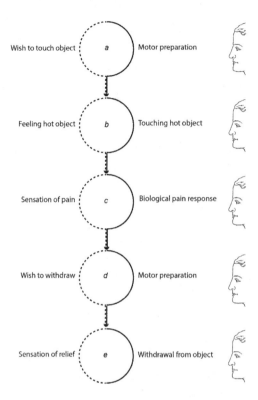

Wish to touch object · a · Motor preparation

Feeling hot object · b · Touching hot object

Sensation of pain · c · Biological pain response

Wish to withdraw · d · Motor preparation

Sensation of relief · e · Withdrawal from object

26. This figure shows a model of the relationship between psychological states and physical events as understood using the psychophysical parallelism framework outlined here. The psychological event of wishing to touch an object (dashed side of circle *a*) is the psychological aspect of an observable motor preparation event in the brain (solid side of circle *a*); these events jointly cause the act of touching the hot object (circle *b*), in turn causing the sensation of pain and the observable physiological response (circles *c* and *d*) resulting in withdrawal from the object (circle *e*). Note the parsimony and the lack of explanatory gaps in this model, compared with that illustrated in figure 25. Note also that this model is much closer to the monist dual-aspect view of mind–matter relations proposed by Spinoza, Leibniz and Fechner, as discussed herein.

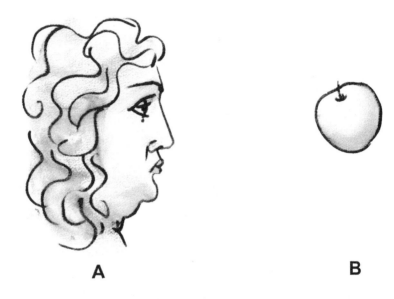

A   **B**

27. In this illustration of the famous story of the apple, Isaac Newton, at **A**, observes an apple, at **B**, that appears to him to be falling—as he famously realised—due to the earth's gravitational attraction. We can imagine what it was like for Newton to experience this event from his own point of view (see the photographs in figure 28). But what was it like—if anything—for the apple to experience being drawn towards the earth from its own point of view, and what—if anything—motivated the earth to draw it? These questions are considered in this section and in part III.4.

III.2   Isaac Newton's apple

One of the most famous stories in the history of science is that of Isaac Newton conceiving the universal nature of gravity when observing an apple fall to the ground. Newton told the story to several people, including his biographer William Stukeley, who recorded a conversation they had in Newton's old age:

> After dinner, the weather being warm, we went out into the garden and drank tea under the shade of some apple trees, only he and myself. Amidst other discourse, he told me he was just in the same situation as when formerly the notion of gravitation came into his mind. "Why should that apple always descend perpendicularly to the ground," thought he to himself, occasion'd by the fall of an apple, as he sat in a contemplative mood. "Why should it not go sideways or upwards, but constantly to the earth's centre? Assuredly, the reason is that the earth draws it."[85]

It is natural to assume that Newton contemplated the falling apple from an extrinsic perspective—that is, from his own point of view—and that he did not consider the experiential nature of the apple when being drawn to earth or the psychological motives of the earth for drawing it. As has been pointed out, such is the conventional viewpoint adopted by scientists when studying the behaviour of material systems. Few people have considered—let alone seriously studied—the possible psychological states experienced by material systems or addressed the questions raised in the caption for figure 27.[86]

---

[85] Stukeley (1752/2004).

[86] Exceptions would include the (primarily) nineteenth century figures who subscribed to the notion of "sensitive matter", such as those mentioned in part I and Ernst Mach, as discussed later in part III.6. See also part III.4 for a discussion of what a falling apple might experience. Newton did hint at the problem, as discussed in I.7.

Although we cannot directly observe the falling apple's psychological states or the earth's motives for attracting it, the framework set out in this treatise allows us to infer those psychological states by measuring the mass and acceleration extrinsically, as set out in part II. We could then, if we wished, explain the observed behaviour by the effect of the earth's Will (which is far greater in magnitude than that of the apple) and the exercising of its Volition to conjoin with the apple which, by exercising its own Volition, affects the earth by displacing it, however slightly, from its preferred state of Repose.

Perhaps due to the continuing influence of Descartes's dualism, many people today would be sceptical about attributing feelings and motives to material systems, and even more so about the value of measuring them. But this scepticism can be countered by the argument given in part I, which is that we ourselves are examples of material systems that have feelings and motives. That our conscious experiences of these feelings and motives are most probably due to the highly complex organisation and behaviour of the organic matter and electrochemical energy in our nervous systems, as also discussed in part I, does not change this basic fact.

It seems odd that Descartes' views on mind–matter relations should predominate today while those of Newton are largely forgotten, despite his greater scientific legacy. Descartes, for example, plainly asserted that when matter "strives"—as he put it—to achieve some motion it is not motivated to do so by any "thought".[87] Newton's view, particularly as expressed in his post-*Principia* works, was quite different. Rather than following Descartes in making a distinction between soul and body (or mind and matter), Newton, like others of his circle, distinguished between "passive" and "active" principles.

---

[87] Descartes (1644/1983). I think most people today would tend to agree with Descartes; Spinoza took a different view (see Rocca, 2021).

Passive principles are inert, wait to be acted upon and resist being acted upon, while active principles have the agency and motive power to effect change. These principles are not, in Newton's thinking, ontologically separate in the way that mind and matter (mostly) were for Descartes—a separation that Newton deemed unintelligible. Rather, it is due to the unity and complementarity of these principles that the world is as it is and behaves as it does.

For Newton the universe is imbued with active principles, which are expressed not only in the forces that cause motion in otherwise passive substances, but also in life, thought, will, volition and ultimately in God himself. It was in response to criticism that he had not explained the *cause* of gravitational attraction in the *Principia* but only calculated its effects that he later attributed this cause to the active principle inherent in God's will.[88] According to Newton's later views:

Life and will are active principles by which we move bodies & thence arise other laws of motion not yet known to us. And since all matter duly formed by generation & nutrition is attended with signs of life, & all things are framed with perfect art and wisdom, & nature does nothing in vain; if there be a universal life, & all space be the sensorium of an immaterial living, thinking, being, who by immediate presence perceives things in it as that which thinks in us perceives their pictures in the brain and whose ideas work more powerfully upon matter than the imagination of a mother works upon an embryo, or that of a man upon his body for promoting health or sickness, the laws of motion arising from life or will may be of universal extent.[89]

---

[88] Concerning the development of Newton's ideas on active and passive principles, God's will as the cause of gravity, and his view that all nature is alive, see McGuire (1968). For Newton's views on mind–body relations, see Iliffe (2017).
[89] Isaac Newton, Query 23 (draft), *Opticks*, 1706. Quoted in McGuire (1968).

Some commentators have seen Newton's descriptions of active principles, like those in the passage above, as foreshadowing what later became known as *energy*; that is, a power that animates the entire universe, having the agency and capacity to cause change in—and even impart life to—otherwise inert or dead matter.[90]

We can now see more clearly the contrast with Descartes' metaphysical worldview. The mental world, for Descartes, sits apart from a material world which consists of purely mechanical motion. Yet the material world is somehow capable of being influenced by the mental one, at least in certain cases. For Newton, meanwhile, the mental world is inextricably 'folded in' to the material world through its active principles that animate the world and give it form by interacting with its passive principles.

Newton's worldview, then, has an affinity with the framework outlined here, which—for experimental and explanatory purposes— treats matter as possessing the general property of Experience, i.e., the psychological parallel of energy transfer or interconversion. The transfer of energy and the parallel transfer of Experience (both of which are 'active') from one portion of matter to another (matter being 'passive') produces an outward change in the motion of the matter (as work is done against resistance) and an inward change in its psychology (as it is put into Conflict) in a way that is not unlike the operation of Newton's active and passive principles.

In this Newtonian context, it is just as plausible to think of the causal power of gravity in terms of the Will of systems to maintain their Repose in accord with their Volition as it is to think of it in terms of forces of attraction that are proportional to distance and mass. If the observable behaviour of a system can be equally well described and predicted using a framework based on the intrinsic properties defined in part II (or some improved version thereof) as it can using the

---

[90] For a discussion on this point, see Coopersmith (2010).

familiar extrinsic properties, then who is to say that the former framework is not as explanatorily valid as the latter?[91]

The contemporary physicist might argue that we can fully account for the behaviour of the falling apple using Newton's laws of gravity and motion, and so there is nothing left to be explained. In response, it can be pointed out that despite great advances in our understanding of gravity since Newton's time, its intrinsic nature remains unknown. We can say with certainty *what* the apple will do but we still cannot answer the question Newton asked himself in his garden, which is *why*. Attributing psychological states and motives to matter—as proposed in this treatise—might well contribute to such an explanation.

Newton may have appreciated this. He was certainly mindful of the fact that by making outward observations of physical behaviour, and even by defining the laws that govern that behaviour, we cannot fully account for the inward aspects of experience. Other 'psychological' powers and laws so far undiscovered but worth searching for, he thought, must be at work:

> We find in ourselves a power of moving our bodies by our thoughts (but the laws of this power we do not know) & see the same power in other living creatures but how this is done & by what laws we do not know. And by this instance & that of gravity it appears that there are other laws of motion (unknown to us) than those which arise from vis inertiæ (unknown to us) which is enough to justify & encourage our search after them. We cannot say that all nature is not alive.[92]

---

[91] Modern physicists can choose from several frameworks to solve problems of mechanics—including Newtonian, Lagrangian, Hamiltonian and quantum—using whichever is most useful for the problem concerned. The framework outlined here may offer another option that is useful for the study of psychophysical systems.

[92] Isaac Newton, Query 23 (draft), *Opticks*, 1706. Quoted in McGuire (1968). It is interesting to note that Newton attributes thought and the power of mental causation to other living creatures, which was something Descartes famously refused to do; he regarded animals, like material systems in general, as soulless machines.

28. This sequence of images is from a video made by the author which shows an apple falling in front of a tree in the garden of Woolsthorpe Manor in England, where Isaac Newton spent his early years and did some of his most important work on optics and gravity. Newton claimed to have been inspired to contemplate the universal nature of gravity by just such an event.

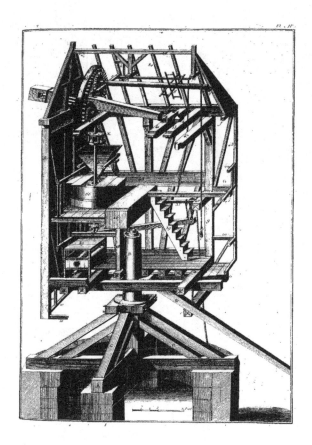

29. An illustration of the workings of a post mill, from Diderot's *Encyclopédie* of 1772. As wind acts on the sails (not shown), its energy is transferred to rotary motion in the wind shaft and in turn transmitted, via the upright shaft and gears, to the stones of the mill. In this process, work is performed on the parts of the mill and on the grain, moving each portion of matter against its Will and imparting to it the property of Experience. The collective Experience of the system while in operation cannot be observed extrinsically, yet by measuring the transfers of energy and the work done, we could precisely calculate what it feels like intrinsically for the mill.

III.3   Gottfried Leibniz's mill

If Isaac Newton's apple is one of the best-known objects in the history of science, then Gottfried Leibniz's mill is one of the most famous in the history of philosophy. Leibniz employed the analogy of the mill in various writings during the early eighteenth century as part of a campaign to refute materialist explanations of mind. In one version of the argument, he says:

> ... perception, and what depends upon it, is inexplicable in terms of mechanical reasons, that is through shapes, size, and motions. If we imagine a machine whose structure makes it think, sense, and have perceptions, we could conceive it enlarged, keeping the same proportions, so that we could enter into it, as one enters a mill. Assuming that, when inspecting its interior, we will find only parts that push one another, and we will never find anything to explain a perception.[93]

Windmills were a common sight in the seventeenth century European landscape. Together with clocks, they were among the most advanced forms of mechanical technology known at the time. Part of the intuitive appeal of Leibniz's analogy is that it invites us—as it must have done for his contemporaries—to think of the brain by comparison to the mill as just another kind of machine made of "shapes, size, and motions", albeit one that is far more complex in behaviour and organisation. Nevertheless, the problem is the same in either case: how can a system composed of numb, dumb and thoughtless matter come to think, "however organised it may be", as Leibniz put it?[94]

---

[93] Gottfried Leibniz, GP: VI, 609/AG: 215. The citations labelled 'GP' are from the collected works published in Leibniz (1875–90/1960). For an insightful discussion of Leibniz's mill argument and its place in his thought, see Lodge (2014).
[94] Ibid. GP: IV, 482/W&F: 16.

For Leibniz, this was not a question of our ignorance about *how* machines like mills or brains might produce minds. Rather, material systems cannot produce minds *in principle* because they are not real or active in the way minds are. Matter in Leibniz's understanding has no agency or power to act—it is passive. Machines like mills and brains, on his account, are mere "aggregates" of insubstantial passive matter; and since machines are made of this 'extended' matter, they are made of something that can be divided infinitely. Being infinitely divisible, this matter has no substance and, therefore, no "real being".[95]

Mind, on the other hand, has real being because it is genuinely unified and indivisible, or at least composed of what he called "simple" substances that are themselves indivisible and, moreover, endowed with the power to change—i.e., they are active. Our experience of being a conscious, perceiving "simple substance" is not only unified but also expresses "a multitude in a unity"; that is, there is something it is like, consciously, to have many mind-like parts that coalesce into one.[96] For machines to be able to think they must have the power to act and to express this multitudinous unity. Being aggregates of passive and insubstantial matter, says Leibniz, they lack both qualities.

In one sense Leibniz's distinction between indivisible mind and divisible matter echoes that made by Descartes, as noted in part I. However, as matter for Leibniz is not really a substance, he is not a substance dualist in the same way as Descartes. Instead, Leibniz's stance on mind–matter relations is often described as a form of parallelism, not dissimilar to that championed before him by Spinoza and later by Fechner. Leibniz's parallelism nevertheless has some unique features, one of the most famous being its dependence on the doctrine of "pre-established harmony".

---

95 Ibid. GP: 96/LA 120.
96 Ibid. GP: VI, 608/AG: 214. See also the reference to Nagel (1974) in note 32.

Unlike Fechner's parallelism, which sees mind and matter as but two aspects of a single identity (he calls this the "identity view"), Leibniz's form holds them to be distinct. But Leibniz and Fechner both deny that any causal interaction occurs between mind and matter, unlike Descartes' dualism, which posits two distinct substances that can causally interact under certain circumstances (as we saw in III.1). In the analogy Leibniz himself used on several occasions to make this point, mind and matter are like two clocks that keep time in perfect harmony—as pre-established by God—and continue to do so without exerting any influence on each other.[97] It is only the precise and reliable coincidence of material and mental events that gives the appearance of one causing the other, says Leibniz.

Whatever one may think about Leibniz's metaphysical worldview, there is a certain irresistible force to his point about the mill that many people who study the neuroscience of consciousness today would have to acknowledge. However minutely we observe a human brain, using whichever of the vast array of ever more powerful imaging tools at our disposal—and even if we describe the brain as a highly sophisticated molecular machine, far beyond any level of complexity that Leibniz could surely have imagined—we still cannot see anything in that organ that we can directly identify as a conscious experience. In this respect, we have made little progress since the time of Leibniz.

But we do not need to appeal to Leibniz's metaphysics to make the key point, which follows logically from the form of parallelism adopted here for the purposes of scientific study. For it is not the insubstantiality of matter, nor the limitations of our imaging tools, nor

---

[97] Leibniz's use of the clock analogy changed subtly over time; see Scott (1997). He later abandoned it altogether after criticism that it was too mechanistic, replacing it with the analogy of two bands of musicians that are oblivious to each other but nevertheless, by following the same notes, produce harmonious music. This musical analogy serves a similar explanatory purpose to the one that was used here in part I.18.

or our ignorance about what to look for that prevents us observing mental states in brains. We are unable *as a matter of principle* to observe from an extrinsic perspective the psychological states that occur from, and only exist from, the intrinsic perspective of the system concerned, whatever its nature or constitution.

As for the reality or substantiality of mind or matter (which Leibniz seems to have denied in the case of matter and Fechner remained neutral about in the case of both), the unobservable aspect of a system is no less or no more real than the observable aspect since they are really the same system, just considered from opposing points of view. Put another way, if one is real then so is the other as they are both one and the same.

In applying this form of psychophysical parallelism to the problem posed by the analogy of the mill, we can treat the mill extrinsically as a physical system composed of portions of matter "that push one another" (as Leibniz puts it) but *at the same time* as a system composed of intrinsic psychological states that occur in parallel and in perfect harmony with the movements of its physical parts. If we put aside Leibniz's anti-materialism and regard these parallel physical and psychological states as two aspects of the *same* system, taken as a whole—as Fechner was to do later in his identity view—then the main problem posed by the mill analogy evaporates.

Using the methods outlined in part II, we can now account for the psychological processes—the 'non-conscious drives'—occurring within the mill as we observe its various parts push and pull, experiencing fluctuating quantities of Distress and Relief of varying levels of Intensity, or Affliction and Delight, as they move, transferring kinetic energy supplied by the wind to perform the work—grinding the grains to flour—that the mill is designed to do. By applying these same principles to the study of nervous systems and analysing their quantities and qualities of Experience as they transfer energy between parts of their organic matter to carry out biophysical work, we may, as shown in part I, explain conscious states in the same way.

Outwardly, the motions we observe in the mill and the brain are determined by physical causes, i.e., interactions between forces and masses, or molecules and cells, which all involve transfers of energy. Inwardly, the experiences of both systems are determined by psychological causes, i.e., transfers of Experience and the wish of each part of both systems to satisfy its Will and Volition. In fact, the framework outlined here accords with Leibniz's view that *only* mental events cause other mental events (endorsing what is sometimes called intra-substantial causality), re-expressed in the form of proposed interactions between psychological properties like Conflict, Will and Volition.[98]

There is another important respect in which Leibniz's outlook relates to the present framework, and indeed to the recent neuroscientific theories of consciousness summarised in part I. His mill analogy was meant to illustrate that thinking is impossible in a machine—however complex its organisation—because the machine, being merely an aggregate of infinitely divisible parts, lacks the unity, or the unity of the multitude, that Leibniz deems essential for perception and consciousness. How can all these discrete portions of 'sensitive' matter coalesce to form one psychological entity?[99]

The explanatory framework outlined here offers a direct response to this challenge. Considered extrinsically, brains are material systems made of many parts, both anatomical and functional, as discussed in part I. As also discussed there, evidence gathered by observing brain behaviour strongly suggests that when highly differentiated neural activity becomes sufficiently integrated a special critical state, or phase, of neural organisation 'blossoms', and the owner of the brain will most likely be consciously experiencing.

---

[98] On Leibniz's theory of intra-substantial causality, see Jorati (2015).
[99] This also relates to the 'combination problem' mentioned in note 34.

When we consider that same neural activity intrinsically, then the same organisation of the brain's psychological properties (properties such as Will, Conflict, Distress and Relief) likewise blossoms into a special phase or critical state—or process—of *conscious* Experience that has overall coherence for the system, again as discussed in part I. We can ask whether this diverse yet coherent state—considered both extrinsically and intrinsically—constitutes the perceptual "multitude in a unity" that Leibniz argued could not arise by combining material parts, like those that make up the mill.

As outlined in part I, there is nothing impossible or even unusual about the emergence of, or the phase transition to, qualitatively different states when the system has sufficient component parts (i.e., differentiation) that are organised and behave in sufficiently complex and coherent ways (i.e., integration); oft-cited examples from the animal world are the exquisite flowing patterns produced by flocking birds or shoals of fish.

It seems very likely that the emergence of a conscious, and indeed self-conscious, phase in the organisation of highly differentiated yet integrated flows of energy and Experience through astronomically complex nervous systems is another example.[100] Although the unified non-conscious Experience of a relatively simple working mill bears little resemblance to the far more complex conscious Experience of a waking human in terms of its richness and resonance, it is hypothesised here that there is no fundamental difference in the psychophysical processes that produce them both.

---

[100] Seager (2022) proposes the emergence of organisational complexity among elements of "mental chemistry" as a possible source of unified consciousness and as a solution to the combination problem highlighted in note 34.

30. A drawing of a post mill with sails scratched into a wall at Woolsthorpe Manor, discovered in 2017 and thought to have been made by a young Isaac Newton while he was living there. Note that the scratched lines have been digitally enhanced for legibility. Post mills were one of most advanced and complex forms of mechanical technology known in Newton's time, and his experience of seeing them in the Lincolnshire landscape may have fed his curiosity about mechanics. This image is provided courtesy of Dr Chris Pickup, who was part of the team that discovered the drawing. (Reproduced by kind permission of the National Trust.)

31. A reproduction of the frontispiece from Émilie du Châtelet's major work on natural philosophy, *Institutions Physique (Foundations of Physics)*, first published in 1740. See the text below for a description.

III.4 Émilie du Châtelet's living force

In the late seventeenth century, Gottfried Leibniz published a series of papers that triggered a prolonged and impassioned dispute about the nature of force. It involved some of the greatest natural philosophers of the age and eventually led to our modern concept of energy.[101]

The dispute was primarily about how to measure of what was then called the "quantity of motion" in collisions between bodies. According to the view set out by René Descartes in his *Principles of Philosophy* and maintained by his followers, this quantity was equivalent to the weight of the body times its speed, which is similar to the modern quantity of momentum (mass × velocity). As much for theological as for scientific reasons, Descartes believed that this quantity is conserved in all interactions between bodies. As he succinctly expressed the principle:

> We must reckon the quantity of motion in two pieces of matter as equal if one moves twice as fast as the other, and this in turn is twice as big as the first.[102]

Leibniz accepted, again on theological grounds but also on metaphysical and logical ones, that something in nature was conserved when bodies interact but denied that this could be fully accounted for

---

[101] Leibniz (1686). For an account of the dispute see Iltis (1971).

[102] Descartes (1644/1983) Ibid. By "big" he meant heavy. On the principle of conservation, he believed that God, in His perfection, endowed the universe with a fixed amount of motion that could be exchanged between bodies but never gained or lost. Newton was later to enshrine the principle of conservation of momentum in his third law of motion, where each action is met with an equal and opposite reaction.

by Descartes' quantity of motion.[103] Unlike Descartes, who rejected any role for "occult forces", Leibniz believed that there was more to the material world than geometrical bodies in motion. He saw activity in nature as being driven by invisible "motive forces" that cause matter to change its position and form. These forces, Leibniz believed, could be measured by their observable effects and, moreover, were conserved in a way that Descartes's quantity of motion was not.

To make his argument, Leibniz asks us to consider two bodies, one having a mass (although he did not use that term) of 1 unit and the other having a mass of 4 units. Raising the first body 4 units in height from the ground requires as much "force", or what we would now call mechanical work, as raising the second body 1 unit in height. As both bodies fall to the ground, Leibniz reasoned by the principle of conservation that they each acquire an amount of "force" equivalent to that needed to raise them. However, using the Cartesian quantity of motion (i.e., following Descartes) in this situation results in an inequality. The lighter body, in travelling four times further than the heavier one, acquires twice the velocity—as Galileo had demonstrated in the early seventeenth century—and so the quantity of motion (mass × speed) of the former (which is 2) is half that of the latter (which is 4).

Leibniz was correct that Descartes' quantity of motion is not what is conserved in this system. Leibniz's conserved quantity, which he came to call the *vis viva* (living force), is essentially what we know today as energy. The work–energy theorem of modern physics states that the work done on a system is equal to the energy transferred into the system. In lifting a body with mass m above the ground to a height h

---

[103] Leibniz also believed that God had designed the world perfectly and so did not need to intervene to keep bodies in motion. Logically, he placed great weight on the Principle of Sufficient Reason. This mandates that everything must have a cause and non-conserved forces that appear from nowhere would violate this principle.

against the force of the earth's gravitational field g we transfer a quantity of gravitational potential energy (GPE) to the body, defined by GPE = mgh. As the body falls to the ground, the gravitational pull of the earth does work on it, accelerating the body and conferring to it a quantity of kinetic energy (KE), defined by KE = $\frac{1}{2}mv^2$, which depends on the body's mass and velocity v. In an ideal system where we can neglect factors such as friction, the GPE acquired on being raised is fully converted to KE as the body falls, thereby conserving the total quantity of energy, or *vis viva*, within the body as Leibniz claimed.

One of the people to whom we are most indebted for our modern understanding of energy is the eighteenth-century scientist and philosopher Émilie du Châtelet. Much like Elisabeth of Bohemia, she was a highly educated and well-connected member of the aristocracy who not only engaged with some of the leading intellectuals of her day but became one of them herself. The image in figure 31 reproduces the frontispiece from her major work on natural philosophy, *Foundations of Physics*, published in 1740.[104] It is thought to show du Châtelet ascending towards Truth personified as a woman on a pedestal of geometrical forms, while du Châtelet is attended by a bevy of muses, the whole scene presided over by portraits of Leibniz, Newton and Descartes. Light emanates from Truth's arm, a source of energy that is metaphorically equated with truth as befits the ideals of the Enlightenment project which was then in full bloom.

Natural philosophers of the time tended to endorse the physical and metaphysical opinions of Isaac Newton if they were British, René Descartes if they were French, or Gottfried Leibniz if they were German. This nationalistic prejudice partly explains why the *vis viva* controversy that Leibniz triggered through his objections to Descartes'

---

[104] du Châtelet (1740). The book was ostensibly written for the benefit of her son's education, but she sent copies to many of the leading intellectuals of her day, which helped to establish her international reputation as a philosopher and scientist.

mechanics—and which pitted Newtonians, Cartesians and Leibnizians against each other—was so prolonged (lasting some 50 years) and so impassioned; people were sometimes unwilling to concede their views for purely scientific reasons.[105]

In contrast, while du Châtelet was born and worked in France, she did not succumb to the nationalistic bias that often coloured the opinions of her contemporaries, instead providing a compelling and conciliatory synthesis of the major competing views in the *vis viva* controversy in her *Foundations*. She was a leading authority on—and advocate for—the ideas of Newton yet presented a persuasive argument in favour of Leibniz's *vis viva* as the conserved quantity that motivates activity in the world, while at the same time updating and clarifying the relevant aspects of Cartesian and Newtonian mechanics.[106]

Part of the reason that du Châtelet's advocacy of Leibniz's living force moved opinion in his favour was that she supported her theological, metaphysical and logical arguments—many of which she inherited from Leibniz—with solid empirical data. She cited, among others, the experiments carried out in the 1720s by Dutch professor Willem 's Gravesande, in which he collided metal balls of different weights and speeds with balls of clay.[107] The conclusion he drew, and the point du Châtelet was making in reporting his results, was that the

---

[105] David Papineau (1977) points out that the dispute, which was partly about the question of whether forces act by contact or at a distance, was also so prolonged because scientists also needed time to assimilate and explore the implications of the new ideas and data that were then emerging.

[106] du Châtelet translated Newton's *Principia* into French in 1749 and added a commentary in which she subtly recast some of Newton's key principles in terms more consistent with Leibnizian metaphysics. It was not published for a decade after her death but remains the authoritative translation. Today, we treat momentum (mass times velocity) and kinetic energy (half of mass times velocity squared) as distinct quantities that are both conserved in interactions.

[107] 's Gravesande (1722). For more recent commentaries, see Hagengruber (2012).

quantity of *vis viva* (kinetic energy) transferred to a body as it collides with another body is proportional to its velocity squared. This is not the same as the quantity of motion used by the Cartesians, now referred to as momentum, which is proportional to its velocity alone.

The difference is evident from the amount of clay displaced by the impact of the metal balls. In the example given by du Châtelet in her *Foundations*, the degree of displacement is greater for a ball of 1 unit of mass travelling at 3 units of velocity (which has a kinetic energy of 4.5 units, as given by the formula $KE = \frac{1}{2}mv^2$) than for a ball of 3 units of mass travelling at 1 unit of velocity (which has a kinetic energy of 1.5 units). Since the amount of clay displaced by the ball on impact is proportional to the energy transferred to it, this experimentally confirms Leibniz's theoretical argument discussed above.

It is often said today that du Châtelet carried out these experiments herself, albeit in a modified form where metal balls are dropped into slabs of clay from different heights and the resulting impressions in the clay are measured to give the results. She may well have done so for she was known to be an experimentalist, but she didn't mention conducting any such experiments herself in the *Foundations*. Like the apocryphal story of the apple falling on Newton's head—an event that he himself never mentioned—the image of an eighteenth-century French noblewoman dropping metal balls into slabs of clay to reveal fundamental truths about nature has embedded itself in the public imagination, even if it may not be true. Nevertheless, both images are potent because they encapsulate foundational scientific ideas.

With respect to du Châtelet's work on living force, what remains potent is the idea that the material world, which was usually presented in Cartesian terms as merely soulless matter in motion, is in fact formed by an invisible agency, *vis viva*, having the power to impress upon and move the matter which receives it and which we ourselves must also embody and express. As we will see in part III.7, du Châtelet's work prefigured the connection between *vis viva* and the energy flows that animate living matter, although it took many decades for this link to become properly established. Natural philosophers of du Châtelet's era were still coming to grips with Leibniz's proposal that alongside the observable world of bodies in motion, there exists an unobservable but no less real and measurable world of forces—much like those Isaac Newton had speculated about and which we might call energy flows or energy transfer today—that endowed the material world with life and, according the physiologist Charles Sherrington who was quoted in part I.3, with mind.

This excursion into the origins of the modern concept of energy offers some important lessons and implications for the explanatory framework outlined here. Due in part to the work of scientists such as Émilie du Châtelet, we are now familiar with thinking about energy as existing in distinct forms, such as kinetic and potential. We have seen this difference expressed already in the example of the bodies that are raised and dropped, and how energy of these different kinds is transferred and converted during mechanical work.

While both kinds of energy are conserved, interconvertible and equivalent to the quantity of work doable or done, they are measured in different ways. On earth, GPE is proportional to a body's distance from the surface of the earth, whereas KE is proportional to the body's velocity (relative to the earth) squared. Why does energy come in these distinct forms that are measured in different ways, despite having so much in common? And what is energy in the first place?[108] These longstanding questions remain open, but we can cast interesting new light on them by considering the nature of energy and energy transfer from the intrinsic perspective.

---

[108] The nature of energy is still debated; see Warren (1982) and Hecht (2019).

Consider what has been called the "simplest problem of Newtonian dynamics", namely the freefall of an apple from a tree to the earth but adding—in reference to the experiments cited above—a slab of clay into which the apple falls.[109] Treating the problem using Newtonian mechanics, we can say that the earth's gravitational field exerts a constant force on the apple that moves it towards the earth at a rate that is the same for any mass. When the apple lands in the clay, it meets an opposing force due to the inertia of the clay–earth system and is accelerated in the reverse direction such that, when observed from a certain frame of reference, it appears to come to rest. The apple also exerts an equal and opposite force on the clay–earth system, so accelerating a portion of the clay's particles away from the centre of the apple and resulting in an impression in the clay that is proportional in size to the force exerted.[110]

We can treat the same problem using the framework of work and energy transfer discussed above. As the apple begins to fall it possesses a quantity of GPE—say, 10 joules—by virtue of its mass, its height above the earth, and the acceleration due to gravity in the vicinity. As it falls, the force of gravity does work on the apple, moving it through the distance of its initial height above the earth to the clay.

---

[109] The physicists and physics educators André Koch Torres Assis and Ricardo Karam introduce the freefall of the apple as the "simplest" two-body problem of dynamics: the branch of mechanics that studies the relationship between forces and motion (Assis & Karam, 2018). But they then go on to point out that it is not really a two-body problem at all, since one cannot completely rule out the influence of all other bodies in the universe on the falling apple. They give further reasons to show that this problem is more complex than is usually supposed and discuss how Newton's awareness of its complexity led him to assume a fixed "absolute space" against which all motion could be measured. As will be discussed in part III.6, this was an assumption that the physicist Ernst Mach later rejected.

[110] It should also be noted that as the apple exerts a force on the earth via the clay as it comes to rest, it accelerates the earth to an (imperceptible) extent that is inversely proportional to the earth's mass.

As this occurs, the GPE initially possessed by the apple is converted into KE, the quantity of which is given by the mass of the apple and its velocity squared. Following the law of energy conservation (and ignoring factors like air resistance and the apple's internal energy), the 10 joules of GPE are fully converted to 10 joules of KE by the time the apple hits the clay, such that GPE = 0 when it is at rest. In other words, the overall quantity of work done by gravity on the apple is equivalent to its initial GPE. During impact, the apple does work on the clay by virtue of the KE it has acquired, transferring that KE to the clay and displacing a portion of it, resulting in an impression that is proportional in size to the quantity of KE in the apple (as was shown in the experiments discussed by du Châtelet).

Accounts such as these are common in physics textbooks and are perfectly adequate if one wants to predict the motion of the apple or the size of the impression it will make in the clay; that is, if one is studying the system from an extrinsic perspective. But the situation is quite different when we consider the intrinsic perspective of the system. If we are interested to know, for example, what the apple feels during its journey from the tree to the clay or what the clay feels when it is displaced by the apple, we need a different account.

In the extrinsic description just given, the apple possesses 10 joules of GPE when at its initial position, which is fully converted to KE by the time it reaches the clay. Here, we face an apparent problem when giving an intrinsic account: the apple retains the same amount of energy throughout its fall, albeit of two different kinds. Yet we also know that *work is being done* on the apple due to the gravitational force, which implies that energy is being *transferred* to the apple during the fall. This should feel unpleasant for the apple due to acquiring Experience, remembering that Experience is defined in part II as the (intrinsic) parallel to the (extrinsic) *transfer of energy*. But according to the standard work–energy account, no energy is being transferred during the fall, only *converted* from one form to another. From this perspective, the apple acquires no Experience, seemingly pointing to an anomaly in the intrinsic account.

This apparent anomaly disappears, however, if we recognise that we are mistaken in attributing the GPE to the apple alone in its initial position. Strictly speaking, the GPE exists due to the relationship between the apple and the earth in terms of mass and distance; it is a property of the *apple–earth system rather than the apple system alone*, although many textbooks fail to point this out. Consequently, the 10 joules of GPE that exist before the apple falls are in fact possessed by the totality of the apple–earth system rather than either of its components individually. We could theoretically calculate the portion of GPE belonging to the apple component by artificially isolating one part of the system from the other, but that portion would be infinitesimally small given the relative masses of the apple and the earth.

As the apple begins to fall, however, an important change occurs. The GPE belonging to the apple–earth system is converted to KE that is possessed by the apple system *alone*. This is because the KE of the apple system is proportional to its velocity, which is measured with respect to the earth where the earth is assumed to be at rest. And since it is only the apple system that has velocity in this frame of reference, it is only the apple system that has KE. On this account, therefore, the GPE in the apple–earth system is both *transferred* to the apple system as it falls and *converted* from GPE to KE so that the total quantity of energy is conserved.[111]

---

[111] Admittedly, it is counterintuitive to treat the apple–earth system and the apple system as distinct systems, especially since the apple is a component of both. Yet it is necessary to do so if our purpose is to accurately track the distribution of the corresponding energy and Experience during the fall. As discussed below, this may become important when we come to analyse the energy transfers in neural systems that parallel conscious Experience.

Translating this into the intrinsic psychological domain, we can say that because the apple system in its initial position possesses effectively zero GPE and zero KE (ignoring internal energy), it has zero Experience and is at Repose. But as it falls it acquires a quantity of Experience due to the KE converted from GPE as energy is transferred from the apple–earth system. Because the change in its Experience is positive, this will have a negative hedonic valence for the apple system and so it will feel 10 units of Distress, measured in Algons (**A**). The apple–earth system, meanwhile, is transferring Experience away from itself during the fall, and so this negative change in its quantity will have a positive hedonic valence that will be felt as 10 units of Relief, measured in Hedons (**H**). Similar calculations can be made for the clay as it is disturbed from its preferred state of Repose when the apple transfers a quantity of Experience to it upon collision.

This situation becomes still more informative if we take another intrinsic psychological property of the system into account, namely the Intensity of the Distress or Relief that is felt during an interaction. As defined in part II, Intensity in Kratons (**K**) is given by the ratio of the Experience ($\mathcal{E}$) of the system to its Will (**T**): i.e., $\mathbf{K} = \mathcal{E} / \mathbf{T}$. Thus, if a system A with a Will of 5 **T** (parallel to its mass) acquires 10 $\mathcal{E}$ of Experience, it feels that experience more intensely—qualitatively speaking—than a system B having a Will of 500 **T** that acquires the same amount of Experience. We can also consider the Intensity in terms of the *direction* of change in Experience, as defined in part II, which in the case of increasing Experience, or Distress, is measured as Affliction ($\mathbf{K_A} = \mathbf{A} / \mathbf{T}$) and in the case of decreasing Experience, or Relief, is measured as Delight ($\mathbf{K_H} = \mathbf{H} / \mathbf{T}$), both of which give the Intensity of the hedonic valence of the system.

Applying these quantities to the case of the falling apple, we can say that the apple–earth system starts with 10 $\mathcal{E}$ of Experience (given that it has 10 joules of GPE) but the Intensity of that Experience—

given by dividing the 10 ε by the Will of the apple–earth system, which is approximately 6×10^{24} **T**, and resulting in an approximate value of 1.67×10^{-24} **K**—is extremely small. Likewise, the Intensity of the Relief, or Delight, felt by the apple–earth system as the apple falls and Experience is transferred out of the system is also extremely small ($\mathbf{K_H} = 10$ **H** $/ 6 \times 10^{24}$ **T**).

The Intensity of the Experience of the apple system itself, however, is quite different. Assuming it has a mass of 0.2 kg (corresponding to a Will of 0.2 **T**), the 10 **A** of Distress it acquires during the fall— as GPE is transferred into it and converted into KE—means it feels an Affliction of $\mathbf{K_A} = 10$ **A** $/ 0.2$ **T**, or 50 $\mathbf{K_A}$, which is far greater than the Delight felt by the apple–earth system. A similar calculation can be made for the Affliction suffered by the clay system as the apple disturbs it from its state of Repose upon impact.

At face value, this account of the psychological states of material systems like apples and slabs of clay may seem rather nonsensical. Why would we be interested in studying such things at all? But considering this seemingly simple problem from the intrinsic perspectives of the apple system and the apple–earth system provides some important lessons that may bear directly on our own experiential nature and on our understanding of nature more widely. This exercise demonstrates that when we study the organisation and behaviour of a material system from its intrinsic perspective, our understanding of the nature of its Experience depends on both the precise delineation of the system (e.g., between the apple system and the apple–earth system) and the respective Intensities of the Experiences of each system, including their relative quantities of Affliction and Delight.

When we come to apply the same principles to the study of the neurobiological processes that parallel our conscious Experiences— that is, to the study of energy transfer processes in biochemical systems at a vast range of spatiotemporal scales, as will be further discussed in part III.7—then we will need to be equally scrupulous in defining the bounds of the systems we are analysing and measuring the qualitative properties of their experiences. The intensity with which someone feels a certain pain, for example, may be determined by relatively small transfers of energy occurring among microscopic biochemical processes in quite specific regions of their nervous system. It will obviously be necessary to identify and analyse these processes accurately if we are to fully understand the cause of the pain and perhaps treat it effectively.[112]

The example of the falling apple also allows us to address the question posed earlier about the (still somewhat mysterious) nature of energy. We commonly think of physical processes from an extrinsic perspective as entailing distinct kinds of energy transfer and conversion, in this case, potential due to position and kinetic due to motion. But this is not so when we consider processes intrinsically; for there is nothing to make a system aware, from its perspective, of whether its Experience is due to the transfer or interconversion of potential or kinetic energy.[113] All that counts from a system's intrinsic perspective—it seems—is the change in the quantity and quality of its Experience and whether it is disturbed from Repose or returning to it. This

---

[112] We get an idea of how Intense the Experience of our organic matter might be due to the ratio of energy transfer to mass at the molecular scale from Nick Lane's book *Transformer* (Lane, 2022). In it he describes the extraordinary effect of the electrical charge on a mitochondrial membrane produced by a certain chemical reaction: "This charge is awesome. It might not sound like much—just 150–200 millivolts—yet the membrane is extremely thin, only 6 nanometres in thickness... This means that the electrical field strength (what you'd experience as a molecule in the membrane) is about 30 million volts per metre, equivalent to a bolt of lightning across every square nanometre of membrane... If you were to iron out all the mitochondrial membranes in the [human] body, so that they were flat, they would cover an area equivalent to about four football pitches—all charged with the power of a bolt of lightning."

[113] Note that, as stated in part II, motion or direction of motion do not play a role in determining the *intrinsic* nature of a system's Experience.

may have important implications and lessons for our wider understanding of reality. For one thing, it suggests that psychological causes operating from the intrinsic perspective of material systems could be explanatorily simpler and more general than those operating from the extrinsic perspective. For example, the inexhaustible variety of observable effects in nature that occur due to transfers of different kinds of energy (thermal, radiant, electrical, magnetic, nuclear, elastic, etc.) might all be extrinsic expressions of one intrinsic causal principle, namely the desire of matter to maintain its preferred state of Repose when disturbed by an influx of Experience.

These implications and lessons point to a deeper understanding of the nature of energy and a more expansive and powerful explanatory framework within which to study natural processes than that provided by the extrinsic perspective alone. It has been instructive to reflect on the scientific contributions made by Émilie du Châtelet and some of those whose work she synthesised, clarified and championed to better appreciate this. Those early scientists found themselves, as we do today, struggling to make sense of things they could not directly observe but which are nevertheless experienced as real through their effects.

With hindsight, it is remarkable that Leibniz, working in the late seventeenth century, was able to use metaphysical, theological and logical reasoning to anticipate the existence of a property of nature that Descartes stoutly denied—i.e., "motive force" or "living force" or energy—which we now take for granted empirically, even if we still do not fully understand what it is. Whether such 'forces' endow matter with mind as well as life, as Leibniz and indeed Newton seemed to suspect, remains to be experimentally studied. But acknowledging that they *may do* is a necessary first step towards finding out.

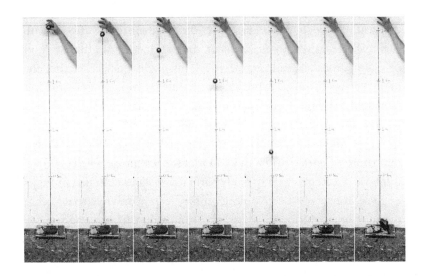

32. A sequence of stills from a video in which the author recreated the experiment often attributed to Émilie du Châtelet which was designed to measure the quantity of 'living force' acquired by a metal ball as it falls into a slab of clay. We obtain a quite different understanding of this event—and the similar event in which an apple falls to the earth—depending on whether we consider it from an extrinsic or intrinsic perspective, as discussed in this section.

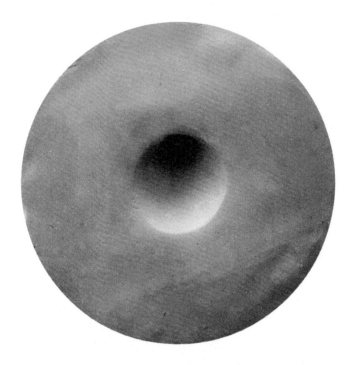

33. The impression made in the clay by the falling ball in the experiment illustrated in figure 32, which is as much evidence of the Experience acquired by the clay, intrinsically, as it is of the work done by the ball, extrinsically. It was impressions such as this that eighteenth-century scientists used to empirically prove the existence of an invisible *vis viva*, or living force, as proposed by Leibniz on theological and metaphysical grounds and later championed by Émilie du Châtelet. As we will now see, later scientists such as Michael Faraday were to discover many more ways in which invisible forces shape the material world.

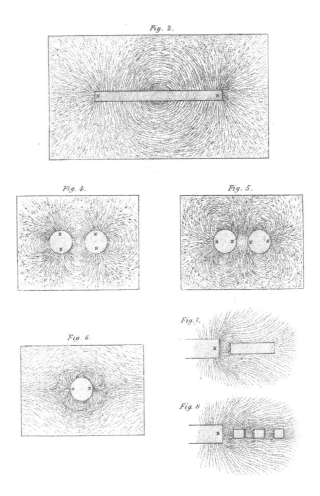

34. This series of drawings was made by the natural philosopher Michael Faraday in the mid-nineteenth century. To create them he sprinkled iron filings over paper treated with wax and then rested it on magnets of different sizes and orientations. The iron filings are forced into a pattern that reveals what Faraday called the "lines of force" that surround the magnets and which are otherwise invisible. By warming the wax, Faraday was able to preserve the patterns produced. From Faraday (1855).

III.5   Michael Faraday's lines of force

During the mid-nineteenth century, the natural philosopher and ex-perimentalist Michael Faraday was working to understand the nature of magnetic and electrical forces. He had already discovered that electricity and magnetism were directly related through his work on induction in the 1830s, which later led to the development of the electric motor.[114] By conducting many meticulously documented experiments in the succeeding decades he became intimately familiar with the observable effects produced by these mysterious magnetic and electrical forces, which he called invisible "powers".

One such effect was vividly illustrated in the images Faraday created by sprinkling iron filings onto sheets of waxed paper and resting them over magnets of various strengths in different positions and orientations (see figure 34). Because iron is strongly influenced by magnetic fields, the particles become arranged into patterns that reveal the spatial distribution of otherwise unseen forces around the magnets. Faraday conceived these distributions of "stress" and "strain" in space as "lines of force" that act on matter in their vicinity through the transfer of magnetic energy or—in the case of electrical lines of force—electrical energy.

Although Faraday did more than any other scientist of the period to experimentally verify the effects of the lines of force that permeate the space around magnetised and electrically charged objects, he was extremely cautious about declaring them to be *physically real*. As noted in the previous section, there had been a long-running and sometimes nationalistically biased debate among natural philosophers in Europe—which had become particularly heated during the period of the

---

[114] Faraday (1832).

vis viva controversy—on the nature of forces, their causal role in producing observable effects, how they could be measured, and whether they could act at a distance without any delay or intervening substance. For example, did gravitational attraction between distant planets—people wondered—occur due to a force of some kind travelling through an invisible intervening medium, or did it occur instantaneously and in the absence of any detectable medium, as it appeared to many at the time?

Historians of science generally acknowledge that by the 1830s, Faraday *had* become convinced of the physical nature of the invisible lines of force—a view which put him at odds with some eminent predecessors such as Henry Cavendish and Siméon Denis Poisson. Faraday's conviction was due in part to the growing body of experimental evidence, much of it gathered by himself, suggesting that electrical and magnetic influences take time to propagate through space. But he waited until 1852 to make a public declaration of his view in a historically significant paper entitled *On the Physical Character of Lines of Magnetic Force*, taking care not to provoke those in the scientific community—of whom there were several—who thought he was becoming deluded by his own imaginings.[115] The fact that the paper referred only to the "physical *character*" rather than the physical *nature* of lines of force in its title may have been a hedge against future refutation of his conviction.

Scientists of the time found it hard to accept Faraday's ideas partly because there was then little conception of what is now referred to in

---

[115] Faraday (1852). One of the precautions taken by Faraday was to publish this paper in a philosophical journal rather than the scientific journal of the Royal Institution as was his usual practice because, as he put it, "it contains so much of a speculative and hypothetical nature." He was perhaps right to be cautious; even in 1855, when Faraday's scientific credentials were well-established, senior colleagues were privately concerned that he was deceiving himself by "attributing an objective existence to his mental images", as reported in Williams (1960).

physics as the 'field'. The twentieth-century physicist Richard Feynman gave a general definition of a field as "any physical quantity which takes on different values at different points in space".[116] While this definition can apply to systems of any sort, as applied to systems of interacting forces the field is understood as a region surrounding an entity—even if that is merely an abstract mathematical 'point centre'—which has measurable properties and can affect or influence another entity, such as a test charge or test particle, in its vicinity.[117] Contemporary physical descriptions of reality rely heavily, and sometimes almost entirely, on the field concept when describing and explaining interactions at all scales, from the subatomic to the cosmic.[118]

Some historians have argued that Faraday was among the first to grasp the field-like nature of the physical world.[119] It is true that he gradually came to understand all of nature, including matter itself, as an expression of invisible "forces" or "powers" acting through space and time, aspects of which we can only glimpse through their effects. This understanding was informed not only by his scientific work but also by his religious beliefs, which led him to hold that there was an underlying unity and interconvertibility between forces such as magnetism, electricity and gravitation—a view that is in sympathy with the

---

[116] Feynman (1964).

[117] McMullin (2002) traces the emergence of the field concept in physics, from its roots in early Greek philosophy, through the work of Kepler, Newton, Boscovich, Kant and Faraday, to James Clerk Maxwell's theoretical summation of Faraday's empirically based field models of electricity and magnetism in what are now known as Maxwell's equations of electromagnetism.

[118] In his book *Waves in an Impossible Sea*, the physicist Matt Strassler (2024) provides a clear account of this widely held contemporary view that reality can be modelled and understood in terms of wave-like fluctuations in fields.

[119] For example, see Williams (1960). Levere (1968), using evidence from private memoranda written by Faraday, argues that Faraday's notion of nature as being composed of fields of force or power "emanating from and around a centre" were influenced by his religious convictions, a theme we will return to later.

unificatory goals of contemporary physics.[120] The extraordinary industry and perseverance he showed in his laboratory work, as well as the intuitive leaps of imagination he made (although only guardedly in public), can be accounted for in part by his drive and determination to understand as much of this unity as he could in the time he had.

Modern physics tends to be circumspect about what physicist Arthur Eddington called the "intrinsic nature" of that which is described by the field. Writing in 1928, he was clear about the limits of the knowledge we derive from physics: we obtain only sets of measurements—what he famously called "a schedule of pointer readings"—that reveal nothing about that which is measured in itself.[121] Nevertheless, the limitation Eddington recognised has not hindered development of physics as a quantitative science. Fields are now regarded as essential mathematical models that enable the behaviour of physical systems to be accurately predicted, irrespective of whatever their intrinsic nature might be. In the case of quantum field theory, physicists are now enabled to measure and predict the behaviour of subatomic events with a degree of accuracy unmatched anywhere else in science.

The development of modern field theory, alongside other advances in physics since the nineteenth century, gradually led to the dissolution of the seventeenth-century view of the world as made of solid, localised and ponderable matter set in motion only by direct

---

[120] Williams (1960).

[121] Eddington (1928). Interestingly, Eddington suggested a view of mind–matter relations not dissimilar to that proposed in this treatise. Why should the intrinsic nature, or the "background", of matter—which is unaccounted for in physics—not be a form of consciousness, he asks, and then ventures: *"There is nothing to prevent the assemblage of atoms constituting a brain from being of itself a thinking object in virtue of that nature which physics leaves undetermined and undeterminable.* If we must embed our schedule of [pointer] readings in some kind of background, at least let us accept the only hint we have received as to the significance of the background—namely that it has a nature capable of manifesting itself as mental activity" (emphasis in the original).

contact. Instead, we have come to view reality as comprising wave-like disturbances in, and interactions between, indefinitely extended fields that propagate energy through time and space—a model inherited partly from Faraday's work on the fundamental role of forces in nature and its subsequent embodiment in Maxwell's equations.[122]

As we have done before, it is instructive to take a scientific concept familiar to those who study the world from an extrinsic perspective and reconsider it from an intrinsic perspective. The earlier discussions of gravity in part III.2 mentioned that observable (extrinsic) physical effects due to mass and attraction can also be conceived in terms of (intrinsic) psychological effects due to Will and Volition. Discussion of these psychological properties has hitherto tended to focus on situations in which forces act on material systems like balls and springs or, in the case of neurobiology, molecules or ions; that is, on material systems we are often tempted to think of as being solid or particulate. But as highlighted through discussing the action of gravity, the psychological properties defined in this treatise can equally be applied to more abstractly conceived systems of fields.

Consider a case where the gravitational field surrounding a system A of a mass m_1 is impinged upon by a test particle B having a smaller mass m_2, with measurable consequences for the motion of B due to work being done on B by the gravitational field of A. Whereas physicists might say that system B 'feels' the influence of A's field due to the force it exerts—without specifying what is meant by 'feel'—we can now be quite specific and even provide (hypothetical) measures of the quantity and quality of the Experience felt by B, and indeed for the reciprocal Experience felt by A. Analogous ideas apply to a test

---

[122] The physicist Freeman Dyson, writing for *Scientific American* in 1953, describes the conceptual transition that physicists underwent, starting with the work of Faraday, from thinking about fields that surround matter to thinking of matter as itself a manifestation of fields.

particle entering the electrical field of a charged system or iron filings that enter the magnetic field of one of Faraday's magnet arrangements (as illustrated in figure 34).[123] Each particle will acquire a certain quantity of Experience as energy is transferred to it by the field it enters, and this Experience will have a valence and Intensity for the affected system that is commensurate with the rate of Experience transfer as well as its Will and Volition.

Applying the present explanatory framework to field phenomena may have its most important implications and practical value in the study of Experience in living systems, and especially in the production of conscious Experience in nervous systems. A growing number of proposals have been made over recent decades about the role played by electromagnetic fields in the production of consciousness in the brain. Some theorists have gone so far as to claim that consciousness is *nothing other* than the activity of the brain's electromagnetic fields.

The biologist Johnjoe McFadden, for example, has made the case in a series of papers published since the early 2000s that synchronised fluctuations in the electromagnetic fields generated by brains, which can be readily detected using imaging tools such as EEG, not only coordinate neural functions but generate a "conscious electromagnetic information field" or more accurately a complex system of interacting fields that encode and integrate information in the brain.[124]

This complex and dynamic information field, which McFadden describes as the "seat of consciousness", is both generated by and influences patterns of neural activity, creating feedback loops that causally

---

[123] In his *Lectures on Physics*, Richard Feynman (1964) defines the quantities inherent in the electromagnetic field "in terms of the forces that are *felt* by a [test] charge" at any point in the field (my emphasis). Physicists assume these quantities exist at that point in the field, he says, whether the test charge is there to 'feel' it or not.

[124] McFadden (2020). Note that the observable EEG signals produced by brain activity are another example of Eddington's "pointer readings".

affect activity patterns at different scales. Evidence in support of this hypothesis is derived from the wealth of neuropsychological data showing close coupling between patterns of fluctuation in the brain's electromagnetic fields and reported conscious states. He also draws on theoretical ideas about the role of complexity—i.e., levels of differentiation and integration—in the brain associated with consciousness, some of which we encountered in part I.[125]

In line with the framework set out here, McFadden argues that we must consider the brain's electromagnetic field from its *intrinsic perspective* in order to understand why—according to his theory—a certain complex integrated electromagnetic field "*feels like* something from the inside", as he puts it.[126] However, he resists any implication that electromagnetic fields are universally conscious. The field created by a toaster, for example, lacks sentience, according to McFadden, as it is insufficiently complex and lacks, as he puts it, the "causal power capable of transferring thoughts to another conscious being".

To my mind, this points to some explanatory limitations of McFadden's theory, which have been recognised in other similar theories.[127] Why should an electromagnetic field "feel like something" when sufficiently complex? How can we relate different properties of the field to specific qualitative conscious states, e.g., to pleasure or displeasure? And how could we measure the consciousness of the fields and make experimentally testable and falsifiable hypotheses and

---

[125] There is an established and fast-developing body of research that highlights the role of electromagnetic fields in the generation of consciousness. Examples include Pockett (2012) and Hales (2014). For a historical review see Jones (2013), and for a recent journal special issue on the topic see Hunt et al. (2024).

[126] Considering the brain's electromagnetic fields from an intrinsic perspective has also been proposed as a possible approach to solving the 'hard problem' of consciousness, as mentioned in the introduction. For a discussion see Jones and Hunt (2023).

[127] See Jones and Hunt (2023) and also McFadden (2023) for a more recent and systematic attempt to show the explanatory superiority of this approach over others.

predictions about them? Moreover, while electromagnetic fields are known to permeate the nervous system and are thought to play a functional role at many different scales, there is no guarantee that they are the only form of biological energy transfer process that does so. There is growing evidence, for example, of the effect of the earth's gravitational field on the brain,[128] while others such as Mae-Wan Ho have argued that organism-wide coherence effects at the quantum scale play a key role in the emergence of consciousness.[129]

The framework outlined in this treatise may be able to help improve the explanatory power of field-based models of consciousness. By applying the framework with its defined quantities and qualities, we can precisely attribute unobservable psychological states of Experience to the organic matter observably affected by the energy transferred to it through, say, electromagnetic fields generated in nervous tissue. Given that the brain generates electromagnetic fields at many scales, from the intracellular level to the global level, and given the vast body of experimental evidence linking the location and behaviour of these fields to measurable psychological responses—as detailed by McFadden and others—we have an opportunity to consider the effects of field-mediated energy transfers from the intrinsic perspective of the nervous systems in which they occur. In fact, I would suggest it *only* becomes possible to explain the *causal* relationship between the effects of electromagnetic (and other) fields and psychological states once we apply the model of psychophysical parallelism introduced in part I.6.

We can also take an important lesson from Michael Faraday's decades-long struggle to understand the nature of magnetic and electrical forces, which had initially seemed distinct and unrelated and could only be detected indirectly by their effects. Arguably, it was because

---

[128] Yokoyama et al. (2021).
[129] Ho (2008). Ho's ideas will be discussed further in part III.7.

of his deeply held (and probably religiously motived) conviction in the essential unity of all forces in nature, and their "powers" to produce the reality we experience, that he persisted in his efforts to uncover evidence of that unity, even when—as happened on more than one occasion—experimental results seemed to show otherwise.

He candidly expressed his views on the nature of these forces in one of his last public lectures, delivered mainly to an audience of children in 1859, where he talked of:

> ...the beautiful laws ... by which we grow, and exist, and enjoy ourselves ... [all of which are] ... effected in consequence of the existence of certain forces, or abilities to do things, or powers, that are so common that nothing can be more so.

Nature, for Faraday, consisted in...

> ...the universal correlation of the physical forces of matter, and their mutual conversion into one another.[130]

It is in keeping with the spirit of Faraday's research that we here assume a fundamental parallelism and unity (even if only for experimental purposes) between the mental and the material, or the psychological and the physical, as expressed in the proposed psychophysical framework. Despite the difficulties this entails—including how to overcome the widely held belief that the mental and material are separate, even incommensurate, domains and that matter does not feel—I suggest that by experimentally studying the psychophysics of material systems we will begin to understand the fundamental unity of nature, as held by Faraday, and the place of mind within it.

---

[130] Faraday (1859).

35. An illustration from Ernst Mach's book *The Analysis of Sensations*, subtitled '*And the Relation of the Physical to the Psychical*', which was first published in 1886. This illustration was based on several pencil sketches made by Mach in the 1870s and '80s in which he depicted the contents of his left-eyed "field of vision" when observing his own body. Compare this illustration to the image in figure 13.

III.6 Ernst Mach's field of vision

The drawing that the Austrian psychologist and physicist Ernst Mach included in his book *The Analysis of Sensations*, which is reproduced in figure 35, is one of the most remarkable images ever made.[131] For one thing, it is the first time in recorded history, to the best of my knowledge, that the egocentric perspective—that most persistent feature of human visual experience—was represented pictorially.[132]

In his drawing, Mach attempted to depict the "field of vision", as he called it, of his left eye, which includes his torso, limbs, eyebrow, nose and moustache. The enigmatic hand holding the pencil poised over his right leg has long excited speculation as to its meaning. Karl Clausberg, a historian of images, suggests it is a hangover from an early attempt on Mach's part to sketch his egocentric perspective; he began by drawing his own hand in the act of drawing but found that it soon filled the page of his sketchbook and left no room for the rest of the view he wanted to depict.[133] Since I had a similar experience when beginning to make egocentric drawings such as the one in figure 36, I can concur with Clausberg's suggestion.

Another reason why Mach's image is so remarkable is that it summarises in pictorial form the essence of his highly influential philosophical and scientific worldview. In effect, it is a pictorial thesis. The reason for including it in *The Analysis of Sensations* was to illustrate the continuity, and indeed the identity, between the "ego" domain (by

---

[131] Mach (1897/1914). First published in German in 1886.

[132] The only other examples I can find of egocentric representations that predate Mach's drawings are some prehistoric figurines and a small sketch made by the artist Albrecht Dürer of his own legs in 1493 (Courtauld Collection, London). Egocentric depictions begin to appear more frequently in European art in the early twentieth century. For more on this topic, see Pepperell (2015).

[133] Clausberg (2007). It's interesting in light of the discussion in the previous section to note Mach's use the of the word "field" to describe the content of visual space.

which he meant the self-aware 'I'), the body domain and the world domain. Mach took this continuity or identity to be a fundamental and "necessary presupposition of all exact research".

As has been pointed out already, there is still a widespread tendency to assume a distinction between these domains, particularly between the mind on the one hand and the body and world on the other, which would probably have disappointed Mach greatly. The efforts he made to refine his position over decades, and the many (often vitriolic) attacks he endured on occasions when he made it public, were all suffered in the service of advancing a profoundly important idea: that there is no essential distinction between mind and matter and that physics, physiology and psychology are but different ways of studying *exactly* the same thing.

Mach arrived at this idea—which he expounded at length in *The Analysis of Sensations*—due in large part to the influence of Gustav Fechner. Reading Fechner's *Elements of Psychophysics* of 1860 had convinced him that the approach which later became known as psychophysical parallelism offered the most efficient scientific framework within which to research mind–matter relations. The guiding principle of this research was...

> ...the principle of the complete parallelism of the psychical and the physical. According to our fundamental conception, which recognises no gulf between the two provinces (the psychical and the physical), this principle is almost a matter of course; but we may also enunciate it ... as a heuristic principle of research.

Mach went even further in suggesting that this principle could be extended to encompass *all* scientific study and beyond, embracing other areas of human endeavour such as philosophy and the arts. His aim was a truly pan-disciplinary and unificatory programme of research.[134]

The essence of Mach's position, as illustrated in the "self-inspection of the Ego" drawing as he called it, is to treat all of reality—including our experiences of our own minds and bodies—as "connections" between the titular "sensations" of the book. The term 'sensations' has caused much confusion as it seemingly implies Mach was an idealist who took reality to be no more than bundles of ideas or sense impressions (as discussed in part I).[135] However, nothing could be further from Mach's view. He repeatedly asserts that his sensations are to be taken as psychological *and* physical, i.e., as psychophysical. All that exists, as far as Mach is concerned, are the *relationships* between bundles of sensations that manifest themselves in our experience. In everyday life, we tend to compartmentalise certain sets of sensations into "nuclei", i.e., objects. For example, we take our own bodies to be unique physical objects that are distinct from other bodies, or our egos to be unique mental objects that exist separately from our bodies. This is a mistake, thinks Mach. Psychophysical objects—or nuclei—do not exist independently but only by virtue of their interconnectivity.

By making the connections rather than what is connected the objects of his analysis, Mach stresses the relational nature of reality and avoids the absolutist tendencies he detected in the Newtonian worldview, with its reliance on concepts of absolute space and time.

---

[134] Staley (2021) provides an interesting survey of the historical and intellectual context within which Mach's ideas developed, and his ambitions for a unified science—an ambition that was not unusual among his contemporaries.

[135] Mach's philosophical views are problematic because, despite his frequent and stout denials that he was an idealist or that he closely followed Fechner's dual-aspectism, many people have subsequently interpreted him in precisely these ways. For a discussion of the critical reception of Mach's philosophy and the claims that he was "phenomenalist" or idealist, see Preston (2021).

This aspect of Mach's thought was to partly yet profoundly impact the development of Albert Einstein's ideas about space-time relativity.[136]

Interestingly, Mach somewhat distances himself from Fechner's dual-aspect parallelism despite claiming lineage from it, although to my mind it is not entirely clear how or why he does so. Mach insists that mind and matter are not "two different aspects of one and the same reality" as he takes Fechner to say, but rather he asserts the complete identity of mind and matter. As he puts it:

> The elements given in experience, whose connection we are investigating, are always the same, and are of only one nature, though they *appear*, according to the nature of the connection, at one moment as physical and at another as psychical elements [my emphasis].

Elsewhere in the book, he nonetheless takes a more Fechnerian line (and one perhaps more consistent with the use of "appear" in the quote above) by saying that the physical and the psychical—or the "outside" and "inside", as he also puts it—are but "one kind of element" which present themselves "according to the *aspect* in which … they are viewed" (my emphasis).

Perhaps Mach was reluctant to fully embrace Fechner's dual-aspect parallelism due to a worry that it carried an unwanted metaphysical implication, namely that the psychological side of the system—the side that cannot be directly experienced by the observer—could be

---

[136] Einstein acknowledged his intellectual debt to Mach on several occasions, as, for example, in the obituary notice he wrote for Mach in *Physikalische Zeitschrift* in 1916. Staley (2021) suggests that Mach's reflections on the physical basis of his own psychological experiences—what Staley calls the "reflections of a *psychophysical* observer" (his emphasis)—may have stimulated Einstein's famous thought experiments about the experience of travelling on a beam of light or being in freefall. These in turn stimulated my reflections on the possible intrinsic experiences of material systems, as discussed in this treatise.

misconstrued as the unknowable 'thing-in-itself'. This was the troublesome metaphysical concept that Mach had initially imbibed by reading Immanuel Kant at the age of 15 but later came to abhor; the idea that there was a separate world of reality beyond what we can observe, he said, was an "absurdity". If this presumption is correct, Mach was being somewhat unfair on Fechner's model, which as far as I can tell implies no such thing and was purposefully free of metaphysical assumptions, just as Mach's model was intended to be.

This digression into the history of nineteenth-century psychophysics, like our earlier excursion into the history of seventeenth- and eighteenth-century mechanics, is important due to its bearing on the scientific framework being outlined here. To begin with, it allows us to better appreciate the intellectual context in which the drawing reproduced in figure 36 was created, and why it serves as an emblem—or pictorial thesis—of Mach's psychophysical approach.

By illustrating the continuity and identity of the self-conscious ego, the body and the world (as noted above), Mach draws our attention to perhaps the only situation we encounter in which the psychophysical parallels converge, or put another way, where the two sides of the moon in the analogy presented in part I.6 are experienced at the same time. Mach's drawing illustrates the fact that when we observe our own bodies, either directly or in a mirror as illustrated in part I.20, we see them from an extrinsic physical perspective while also experiencing them from an intrinsic psychological perspective.

To stress this important point, consider an act of self-observation in the context of the psychophysical framework outlined in this treatise. When we observe ourselves touching our own bodies, we directly experience the feelings or sensations that occur when the psychological properties (as defined in part II) in the matter of our nervous systems are sufficiently organised (i.e., differentiated and integrated, as discussed in part I). By intrinsically experiencing the psychological states occurring in what is, *at the same time,* an extrinsically observed physical object, we are proving that material systems can feel, and so vindicating the psychophysical approach proposed here.

Some might object, "But not *all* material systems can feel!" Attributing feelings to material systems may be fine when they are healthy awake animals with nervous systems that are organised and behaving in complex ways, but not when they are simple mechanical systems like balls and springs or chemical ones like molecules and ions.

Mach would not have been among those objectors. While he is less renowned as a panpsychist or hylozoist, his perspective was aligned with those of predecessors and peers who did hold such views.[137] The statement he made about the unity of matter and mind in his other most influential book, *The Science of Mechanics,* was noted in part I. And he was hurt by the personal criticism he received for saying—as scientists like Ernst Haeckel had done, as also noted in part I—that "sensitivity is a general attribute of matter, more general than mobility".[138]

Indeed, the main purpose of *The Analysis of Sensations* and the egocentric drawing it contains is to argue that *only* by acknowledging the physical nature of that which we commonly take to be psychological—and vice versa—can we achieve a full scientific understanding of nature itself. Though I do not say he would have endorsed the specific proposals made here about the psychological properties of matter, Mach was undoubtedly attempting to realise the same goal: to bring physics and psychology together within a unified explanatory framework, just as Fechner had endeavoured to do before him.

---

[137] In the preface to the third edition of *The Analysis of Sensations,* Mach cites the philosopher and mathematician W. K. Clifford, whose hylozoic theory of "mind-stuff" was noted in part I, calling him a "writer with an extremely close affinity to myself". And in a statement that echoes the intuition discussed above in note 3, Mach also said that "we ourselves and all our thoughts are also part of nature" (Mach, 1883/1960).
[138] Mach as quoted in Heidelberger (2004). See also note 74.

36. *Self-drawn IV*, 2012. Gouache and pencil on paper, 70 × 40 cm (original in colour).

37. *Life worlds*, 2020. Mixed media in petri dish, 6× 6 cm (original in colour). This artwork was made by combining various immiscible materials—mainly resins, oil paints and water-based inks—in a petri dish and allowing them to interact over time. The thermal, kinetic and chemical forms of energy possessed by the various materials tend towards a minimum as they interact, resulting in the spontaneous emergence of spherical forms that are reminiscent of planets or microscopic lifeforms. The image represents the lowest energy or highest (thermodynamic) entropic state of the system as it reaches (near) equilibrium.

III.7   Erwin Schrödinger's negentropy

A book need not be lengthy to have a deep and lasting impact, and the scientific and philosophical impact of Erwin Schrödinger's book *What is Life?* based on a series of lectures he gave in 1943, is inversely proportional to its 90-page length.[139] Schrödinger—a fellow physicist and countryman of Ernst Mach's—is mainly noted for his contributions to the development of quantum physics in the early twentieth century. But his speculations in the book about the chemical basis of heredity, and in particular his suggestion that genetic information is embodied in the "code-script" carried by a complex form of crystalline matter, are claimed to have inspired the research programme that led to the discovery of the helical structure of DNA in the 1950s.

As well as achieving prominence due to the apparent prescience of these speculations, *What is Life?* has attracted interest and aroused debate due to another topic Schrödinger discussed, namely the role of entropy in the generation and maintenance of living systems. He was one of the first to articulate the idea—now widely shared—that there is a direct but non-obvious connection between living processes and the notoriously difficult to understand Second Law of Thermodynamics. This law, which we encountered briefly in part II, was initially formulated in the mid-nineteenth century by scientists studying the nature of heat and its relationship with the mechanical work extracted from engines. It mandates that heat cannot pass spontaneously from a cooler (lower-energy) part of a system to a hotter (higher-energy) part of a system, otherwise it would permit more work to be obtained from an engine than the amount of energy supplied to run it. And since energy is the capacity to do work, this in turn would contravene another cardinal principle of physics: the conservation of energy.

---

[139] Schrödinger (1944).

In this thermodynamic context, entropy is a measure of the extent to which a system's energy becomes more dispersed, or "spread out", and consequently less 'free' to do useful work as it is 'consumed' during an 'irreversible cycle' of the engine.[140] An irreversible cycle is one in which energy in the system is irrecoverably lost or dissipated. Since most natural processes in which work is performed are irreversible, the Second Law is often invoked to account for the tendency of the universe to move towards thermodynamic equilibrium, or stasis, where energy is most widely and uniformly distributed.

As a physicist, Schrödinger's primary objective in *What is Life?* was to understand how the processes that drive living organisms can be accounted for by foundational laws of physics and chemistry, including the Second Law. At the time he was writing, the origins and nature of life—like the origins and nature of mind—were largely impervious to scientific understanding, and they remain so today. In seeking an explanation, Schrödinger began with those fundamental features of reality on which all processes in nature depend, namely, the atomic structure of matter and its energy-driven behaviour.

And here he saw a major problem. The random behaviour of matter at the microscopic scale, he said, ostensibly prevents the emergence of highly organised structures at the macroscopic scale—structures like living organisms—because "orderly" arrangements of microscopic particles are more likely to disintegrate, or to become what

---

[140] According to this definition of entropy as formulated by Rudolf Clausius, with heat-driven engines (and many other devices that generate work, such as water turbines), it is the *difference* between the level of energy in one part of the system and another that provides the free energy necessary to perform useful work. It is the tendency of these energy differences to seek equilibrium that gives the motive power of the engine which is harnessed to do work. Entropy in this context can be regarded as a measure of the extent to which this energy difference and its motive power is 'lost' to the system as work is extracted and the energy in the system becomes more uniformly distributed. See also Lambert (2002) on entropy as "spread out" energy.

he calls "disordered", than to remain orderly. And as they become more disorderly, so their entropy increases.[141] Here, entropy takes on a somewhat different meaning than in its classical thermodynamic context: *statistical* entropy, rather than being concerned with only the distribution of *energy* in a system, also measures the distribution of *matter*, and in particular the way particles of matter tend to arrange themselves in more statistically probable configurations, where possible, depending on the number of degrees of freedom available to them.

For example, consider an apparently orderly looking portion of matter we might observe at room temperature—say, a glass of water. The countless molecules that make up the liquid are incessantly and randomly moving due to the kinetic energy they possess. As they do so, those particles at the surface of the liquid interact with the gas particles of the surrounding air, which are also incessantly and randomly moving due to their kinetic energy. Over time, the division of the system into separate regions of water and air, or the "orderliness" of the system as Schrödinger would call it, disintegrates into "disorder" as the respective molecules become randomly and imperceptibly mixed as the water evaporates. Eventually, when the water has evaporated completely, the water–air system reaches equilibrated uniformity. This "disordered" state (which also entails a phase transition of the water) is favoured by statistical entropy since it is vastly more likely to occur—given all the possible configurations of the water–air system—than the relatively unlikely state in which the water molecules are all concentrated in the glass.

How is it, asks Schrödinger, that this microscopic tendency towards increasing statistical entropy that occurs in all matter does not

---

[141] I show the words "orderly" and "disorderly" in quote marks because they are used by Schrödinger. But doubts have since been expressed about whether it is appropriate to refer to entropy as a measure of order and disorder, and not only because of the subjectivity these terms imply. For a discussion see Lambert (2002).

prevent incredibly large numbers of atoms becoming highly organised, under certain circumstances, and acting in lawful and orderly ways as they do when arranged as living organisms, or even as brains that are capable of conceiving of atoms? As he puts it:

> Life seems to be an orderly and lawful behaviour of matter, not based exclusively on its tendency to go over from order to disorder, but based partly on existing order that is kept up.

It is because living systems are organised in such a way that they can "suck in" orderliness from their environment, Schrödinger goes on to argue, that they not only compensate for the disorderly tendencies of their matter at the atomic and molecular scales but can 'keep up' ever more elaborate and orderly macroscopic structures as life evolves over time. But what is this environmental "orderliness" that life uses to build and maintain itself?

To address this question, Schrödinger introduces a term that provoked debate at the time he presented it and has done so ever since. If all living systems are subject to a disintegrative tendency to increase their entropy, whether of the thermodynamic or statistical kind, then they must survive and indeed flourish on a diet of 'reverse entropy', or what Schrödinger called "negative entropy" or "negentropy". Negentropy provides the impetus towards orderliness and structure in nature by mitigating positive entropy.

Schrödinger defines negentropy as the reciprocal, $1/D$, of the disorder of the system, D. The quantity of disorder, D, is a term in the standard equation used to quantify statistical entropy, which is:

$$\text{entropy} = k_{\text{B}} \times \ln D$$

One way to interpret this formula is to say that the quantity of entropy in a system is related to the amount of kinetic energy of each particle that is transferred into or out of the system, proportional to the average kinetic energy of all its particles (this being the relationship measured by the k_B term in the equation, known as the Boltzmann constant), multiplied by the natural logarithm of the number of different ways in which the particles of matter in the system can be arranged (a number represented by the D term in the equation). Put more intuitively, the entropy of the system increases (although at a decreasing rate) as more kinetic energy is transferred to it and the number of possible ways in which its particles can be arranged increases. Conversely, the entropy of the system decreases as kinetic energy is transferred out of it and the number of possible ways in which its particles can be arranged decreases.

On this basis, a system that is being heated will undergo an increase in its entropy as its particles become more excited and act against each other to distribute themselves more widely in the available space, so achieving the more probable distribution. For instance, the glass of water in our earlier example would evaporate more quickly (to attain the statistically most probably equilibrium state) if the temperature of the system was increased. Meanwhile, a system that is being cooled will tend towards lower entropy because less energy is available to the system; its particles will divest energy and 'pack' themselves into more regular and concentrated arrangements, having fewer degrees of freedom available to them.

Negentropy as defined by Schrödinger is not simply a measure of decreasing entropy, however; it is not just a mark of a system having *less disorder*. On that basis, it would be enough to merely cool a system down for it to spring into orderly life. Negentropy has the more positive connotation of (actively) *increasing order* in the system, where the system—and Schrödinger is thinking primarily about living systems—can extract or "suck in" orderliness from its environment, mainly in the form of sunlight or food, to generate and sustain its own organisational complexity.

Although we might intuitively understand what he means by "orderliness" in this context, Schrödinger is less clear about what orderliness means quantitatively. And it is this aspect of his proposal about the physical basis of life that has led to confusion and debate in subsequent decades. In fact, even at the time he first presented the term 'negentropy' (which he attributes to one of the originators of statistical entropy, Ludwig Boltzmann), it met with resistance from colleagues who questioned its scientific relevance. In an oft commented upon note to the relevant chapter in *What is Life?* Schrödinger expresses reservations about having used the term, saying that 'free energy' would have been the more appropriate term in the physics context. The free energy of a system, as noted above, is that part of a system's store of energy that is free to do useful work, in contrast to the 'bound' energy that cannot be exploited to do useful work.

We might question whether the acquisition of free energy is sufficient in itself to account for the emergence of complex living systems, because one likely effect of transferring such energy to a system will be to heat it up and so produce more entropy rather than less. Schrödinger points out that in warm-blooded creatures such as us, the heat derived from free energy driven metabolic processes might be necessary to speed up the chemical reactions that sustain life. But this still does not directly address the question of why, in living systems, the acquisition of free energy leads to increasing organisational complexity, or "orderliness", over evolutionary time.

One of the many people to address this challenge during subsequent decades was the geneticist and biologist Mae-Wan Ho. In a series of scientific papers and books produced in the 1990s and early 2000s, she developed a novel set of arguments about the relationship between energy, entropy and life that have implications for many areas of science, including our understanding of the biological nature of consciousness.[142]

A key aspect of her arguments was to stress that biological systems do more than import negentropy or free energy from the environment in the form of light and food, and having used it to do biophysical work, increase the entropy of the environment by expelling or dissipating heat. In other words, they do more than rely on energy transfer or flow. Biological systems also *store* energy, and in fact much of the structure and operation of biological systems at all scales of organisation—from the microscopic to the macroscopic—depends on the way in which energy is stored over relatively long periods of time rather than just how it is consumed and dissipated back to the environment as increased thermodynamical and statistical entropy.

Examples of energy storage abound in living systems, insists Ho, such as in the membrane voltage potentials in neurons discussed in part I. These electrical potentials, which are due to differences in the amounts of positively and negatively charged particles either side of the membrane, create a spring-like tension that keeps the system in a 'steady state' away from thermodynamic equilibrium, as discussed in part I, and hold the cell poised and ready for 'action'. Energy storage processes, according to Ho, occur in living systems at a vast range of spatiotemporal scales: for example, at one end of the spectrum, the strain energy stored in protein molecules occurs within a spatial extent of 10^{-9} to 10^{-8} metres and at a timescale of between 10^{-9} to 10^{-8} seconds, while at the other end of the spectrum, some living energy storage processes operate over metre and decadal scales.[143]

Yet stored energy in biological systems is not 'fixed' or static in the way it is in maximal entropy systems that reach thermodynamic equilibrium. In organisms, stored energy can be continuously *mobilised* to

---

[142] Ho (1994); Ho (2008).
[143] Ho (2008).

perform biophysical work—such as when a neuron fires an action potential. Moreover, the mobilisation of stored energy at one scale is dynamically coupled to the mobilisation of stored energy at other scales, from the quantum microscopic to the classical macroscopic. The entire living organism, for Ho, can be thought of as an intricately organised structure that is storing and mobilising energy through efficient coupling between its parts at multiple scales, often doing so in ways that are so exquisitely tuned and efficient that they produce almost no change in entropy, i.e., they 'waste' almost no free energy, and in some cases are reversible, so producing no change in entropy at all. Negentropy, as Ho defines it, is "stored mobilisable energy in a space-time structured (organised) system".[144]

The work of Schrödinger and Ho, alongside that of other biologists, physicists and chemists who follow a similar vein, teaches us that there is an intimate relationship between energy flow, storage and mobilisation in material systems, the production of entropy change as work is performed, and the emergence of highly structured and orderly systems such as organisms that defy the tendency towards organisational disintegration which we observe at the microscopic level of reality. But perhaps because this relationship is very subtle (as Mae-Wan Ho describes it), and very complex (in a way that I have hardly done justice to here), and even seemingly paradoxical (how can processes that seem to destroy order also create order?), we still lack an understanding of the biochemical processes that generate and sustain life and how they are governed by fundamental laws of physics.

Here I want to raise the possibility that further understanding could come from considering the physical, chemical and biological processes that play a role in generating life from their *intrinsic* perspectives. As has been noted several times in this treatise, physical systems are traditionally studied in science from their extrinsic perspectives

---

[144] Ho (1994).

alone. This is equally true for the work on the origins of life and negentropy cited so far, obviating any consideration of psychological motives and desires. Yet the behaviours of living systems are often governed by psychological motives and desires, the most obvious examples being us. If we assume—as has been done throughout this treatise—that *all* material systems have a capacity for acquiring psychological properties and, moreover, that those properties play an intrinsically causal role in the behaviour of the systems, then if we do the relevant experiments we may find that they also play a part in explaining the origins and behaviours of living systems.

For example, if we refer to the conjectures and principles presented in part II we see that Principle 2 states:

> Every system—as observed physically—tends, where possible, to minimise the energy it acquires and maximise the entropy it produces; considered psychologically, it endeavours to minimise its Distress and maximise its Relief.

At first sight, this statement seems to be at odds with the biological principles outlined in this section so far and with the apparent behaviour of living systems in general. Surely organisms are paradigmatic examples of energy-acquiring systems that don't just passively absorb energy in many forms from the environment but actively work to source and mobilise it? The fact that the existence of life depends on what Schrödinger called "sucking in" orderliness, negentropy or free energy from its environment—or what Ho calls the storage and mobilisation of energy "trapped" in sunlight or food—suggests that organisms tend to *maximise* their energy acquisition, within suitable bounds, rather than minimise it. Likewise, the proposal that systems tend to maximise their entropy production seems to run counter to Ho's argument that certain processes occurring in biological systems are so efficient and microscopic that they produce little or no entropy, where entropy production is understood in the sense of free energy dissipation. But consider the *psychological* states of systems also referred to in Principle 2, as stated above. From this perspective, the motives and behaviours of the organism and the operation of the biophysical processes that sustain it look quite different.

The first point to note is that many material systems, not only living ones, regularly and passively absorb energy from the environment, as for example when a rock absorbs radiant energy from the sun. In this case, the main effect of energy absorption will be to heat the matter of the rock, increasing the rate of energy transfer between its atoms and thereby increasing its statistical entropy. From its intrinsic perspective, the rock will experience Distress as it acquires energy. Its atoms or molecules will act excitedly to try and minimise their Distress by moving apart from each other and seeking greater Repose. But short of being heated to melting point they are relatively constrained in their options for movement due to the various forces, or Conflicts, acting upon them within the system and have no option but to suffer the Distress until the heat is dissipated and they can return to Repose.

In different kinds of non-living material systems, however, passive energy absorption can produce remarkably different effects due to the physical properties of the matter. The pattern in the Chladni figure shown in figure 19, for example, is produced when particles of sand are spread on a metal plate that is vibrated by 'playing' it with a bow. Depending on the bowing action and the material properties of the plate and the sand, symmetrical patterns will spontaneously form in the sand, such as the one shown in the aforementioned figure. Similar effects were produced by supplying acoustic energy to a great variety of liquids and particulate materials in experimental work by the Swiss doctor Hans Jenny in the 1960s, some of them quite astonishing in their complexity and lifelikeness. He coined the term 'cymatics' for this phenomenon in which matter takes on often intricate dynamical forms when subjected to energy influxes within certain critical ranges.[145] These effects are due to the mechanical properties of the matter concerned, and particularly the way it vibrates or resonates at different frequencies in sympathy with the level of energy absorbed, with each frequency producing a different resonant pattern.

In her book *The Rainbow and the Worm*, Mae-Wan Ho discusses several examples of well-known thermal and chemical processes that produce similar effects.[146] These include the patterns known as Rayleigh–Bénard convection cells that can form in a container of liquid that is heated from below, and the Belousov–Zhabotinsky reaction in which mixing certain chemicals produces colourful and oscillating spiral patterns. Both are examples of the class of phenomena known as 'dissipative structures', as introduced in part II.

Dissipative structures tend to self-organise to most rapidly and efficiency minimise the energy they absorb, either from the environment or from that released internally by chemical reaction, while also exporting or dissipating less organised energy (i.e., heat). In doing so they achieve a dynamism that is far from equilibrium (as discussed in part I with respect to neurons) where unbalanced forces continue to do work on parts of the system, but the system is maintained in a steady configuration—a configuration that is formed by the influx of free energy and in turn determines the flow of that energy through the system. Much of Ho's 2008 book is concerned with showing how natural systems, especially those that are dissipative structures, can increase in coherence, organisation and complexity and achieve dynamic

---

[145] Jenny (2007). There is a remarkable documentary about Jenny's work, titled 'Cymatics: Bringing matter to life with sound', that can be seen on YouTube. As the film's subtitle implies, it is hard to believe that some of the structures shown in the film, which are highly complex and animated, are not alive. Yet they are not; they are merely examples of the spontaneous patterns and behaviours that can arise when suitable material systems are supplied with suitable kinds and quantities of energy.

[146] Ho (2008).

far-from-equilibrium states, even to the point where they come to life, by ever more elaborate implementations of energy flow, storage and mobilisation processes and their exquisitely organised coupling at many spatiotemporal scales.

Now let us imagine what it might be like for a typical dissipative structure to experience these processes from its own intrinsic perspective. Following an influx of energy, the system considered at the macroscopic level will experience a quantity of Distress that is proportional to the quantities of Experience transferred to it, the Nolition it undergoes and its quantity of Will. If we consider the system in terms of its microscopic components, these macroscopic (psychological) quantities will represent the sum of the Distress, Nolition and Will of each of its component parts.

The effect of this influx will be unpleasant for the overall system and its parts, so it will endeavour to organise itself into a condition that is less unpleasant and closer to its preferred state of Repose. This it will do to the greatest extent possible given its constitution and circumstances. Extrinsically, this will be the condition that we observe to have a degree of organisation or orderliness in terms of the patterns or structures it forms; what we observe extrinsically as the formation of (dissipative) structures at dynamic equilibrium is the effect of the system minimising its Distress intrinsically.

But there is another factor to consider. Given the system reconfigures itself to minimise its Distress, it can also experience a certain amount of Relief, which is pleasurable for the system and its parts. The two complementary motivations of the system are thus being satisfied simultaneously as it strives to reach its preferred state of Repose, and again we observe this extrinsically in the dynamic organisation and behaviour of the system.

This thought exercise sheds new light on the concept of negentropy, particularly when we begin to think about living organisms as examples of highly organised dissipative systems that absorb energy from the environment, storing and mobilising it in the way Ho describes, and then increasing the entropy content of the environment by expelling less organised matter and energy. Now we can see that negentropy entails two complementary drivers of the system's organisation and behaviour: the desire to minimise Distress and the desire to maximise Relief. It may be true that "nature abhors a gradient", to use the phrase coined in a seminal 1989 paper by Eric Schneider and James Kay to encapsulate the tendency for natural systems to spontaneously eliminate energy differences across a distance.[147] But nature also *loves a gradient descent.*

On this account, negentropy is more than the (somewhat ill-defined) "orderliness" that Schrödinger posited as the driver of living organisms, or even the free energy that systems can absorb from the environment. And nor it is just the "way energy is trapped, stored and mobilised in the living system", as Ho expresses it, where such processes generate and sustain ever more coherent and complex organisation and behaviour of matter and energy, as she shows in her book. By considering the intrinsic perspective of the systems, and the causal role played by (intrinsic) psychological properties, we can now also understand negentropy as a process whereby a system *optimises its psychological responses to influxes of Experience.* It is the *interaction* between the material system (and its parts) and the energy it acquires or that flows through it, as observed extrinsically, which produces its observable organisation and behaviour; it is this same interaction which, considered intrinsically, produces its psychological response to its own organisation and in turn organises its own psychological responses.

---

147 Schneider and Kay (1989). For a book-length discussion of these ideas see Schneider and Sagan (2005).

This suggests a tentative proposal about the role of negentropy—defined intrinsically as *the process by which a system organises itself and behaves to achieve optimal psychological states*—in the emergence of living systems. The examples of short-lived thermally and chemically driven dissipative systems discussed so far, such as Rayleigh–Bénard cells and the Belousov–Zhabotinsky reaction, remain in existence only as long as they are supplied with the right kind and quantity of free energy. Living systems can maintain their structures over long timescales, however, by virtue of their ability to mobilise internally stored energy, as discussed by Ho, provided they can continually replenish their energy stores by taking in nourishment from outside their boundaries.

To be able to acquire, store and mobilise these sources of energy, the organisation and behaviour of living systems must be vastly more complex—i.e., their parts must be much more highly differentiated and integrated—than the simple minerals, liquids or cocktails of chemicals described above, as Ho again discusses in depth. Organisms must have suitable apparatus for sensing energy sources, moving towards them (or away from sources of harm), and for ingesting, digesting, storing and mobilising them to act in pursuit of their life goals.

Besides their respective degrees of complexity, we can ask, what else differentiates living systems from non-living ones, given that both depend on the same physical laws and processes? One possibility that may be worth considering, following what has been outlined here, is that living systems are structured and behave in such a way that they *maximise their opportunities for having positively valenced experiences on average over time.* Putting the same idea in extrinsic terms, they maximise their opportunities for undergoing gradient descent. Imagine a simple living system that can seek out sources of free energy in its environment and absorb, store and mobilise that energy in pursuit of its life goals. Based on the proposals made here the processes of energy transfer among the material parts of the system are paralleled by intrinsic Experiences of different quantities and qualities—as defined in part II—although they will be non-conscious Experiences in this case, given their relative simplicity in terms of the degree of differentiation and integration. Still, some Experiences will be positively valenced for the system and others will be negatively valenced; some will be of greater Intensity and others of lesser Intensity, depending on the nature of the parallel biochemical processes occurring in the organism. Relatedly, some Experiences may be brief and localised to small regions of the system while others may be of longer duration and more widespread.

From the psychological perspective, it is obviously beneficial for the system to be organised and to behave in such a way that negatively valenced Experience is minimised and positively valenced Experience is maximised, within its bounds and over its lifetime.[148] Unlike the spring system discussed in part I, where there is a symmetry between the Distress and Relief it experiences when it is stretched and relaxed, a more complex system may self-organise and behave in a way that breaks symmetry (through forming dissipative structures) in pursuit of greater positively than negatively valenced Experiences.

Although this idea might initially seem rather speculative and abstract, it can be presented in the form of an empirical question that can be readily answered by experiment: *Does an energy-absorbing system having sufficient degrees of freedom to configure itself do so in a way that maximises its positively valenced Experience and minimises its negatively valenced Experience?* Put more simply, does it become a 'pleasure-seeking' system? One can readily think of experiments that could try to answer this question in simulated systems, and even in biological systems directly

---

[148] Note that Ho's *minimisation* of entropy principle depends on highly efficient quantum-scale energy transfers which are, in some cases, reversible. However, there are many other irreversible processes occurring in livings systems that tend to maximise the *rate* of entropy production overall. See Endres (2017) and Swenson (2023) for discussions of the so-called Maximum Entropy Production Principle.

if one can measure the relevant transfers of energy with sufficient precision and analyse them using the psychological properties proposed in part II. The answer to the empirical question would be affirmative if more Relief than Distress is produced in the systems.[149]

The idea that a system might break the symmetry that we see in the spring example—that is, between the amount of energy transferred into it (which is experienced as Distress) and out of it (which is experienced as Relief)—raises the important issue of energy conservation. In the case of the ideal spring, the quantity of energy put in, stored and transferred out of the spring remains constant throughout, even though it is converted from the kinetic to the potential form and back. But how can a more complex system experience a greater quantity of Relief that it can afford from its budget of Distress?

There are several ways that this may be possible, including through the neural temporal coding mechanisms of valence discussed in part I.13. Another possibility is suggested by the thought experiment on the falling apple in part III.4, where we saw that even though the total quantity of energy was conserved throughout the fall, the way it was distributed, transferred and converted between different systems (the apple system and the apple–earth system), and the way those systems felt the valence of that Experience with greater or lesser Intensity depending on their quantity of Will, led to a significant difference in the *quality* of the Experience between the two systems. Therefore, it is not only the *absolute quantity* of Experience transfer (experienced as Distress or Relief depending on the direction) that affects the psychology of the system but also the *relative quality* of that Experience and its distribution through the system over space and time.[150] This suggests

---

[149] For simulated systems I have in mind experiments conducted by Jeremy England and colleagues, such as in Kachman et al. (2017) and as cited earlier in part II.5.4.

[150] Another way asymmetry of valence might occur can be illustrated with the simple spring system and the quantities of Shock ($\dot{A}$) and Joy ($\dot{H}$) as defined in part II. If the

that when studying the 'psychological evolution' of living systems, we might find patterns of organisation and behaviour that exploit this principle to maximise pleasure over time without violating energy conservation.[151]

Besides its speculations about the chemical basis of heredity and its proposals about negentropy, *What is Life?* is also notable for its ruminations on the origins and nature of consciousness. Under the influence of Ernst Mach and the Vedanta philosophical tradition of India that had helped shape his thinking, Schrödinger advocates a form of plural-aspect monism that has some affinity with the framework of mind–matter relations presented here.[152] In spite of appearances to the contrary, according to Schrödinger, all minds and all things are one. The "illusion" of many minds, or of the distinction between mind and world or between objects in the world, is due only to the fact, he claims, that we view the same thing from many different perspectives. He illustrates this with the example of Gauri Shankar and Mount Everest, which both appear to be the highest mountain in the Himalayas when viewed from different valleys.

In an extended and more considered attempt to grapple with the nature of consciousness in his essay *Mind and Matter*,[153] Schrödinger

---

spring is stretched very slowly, say over 100 seconds, as it acquires 10 units of Experience and then released so that it relaxes almost instantaneously, say in 0.1 seconds, the quantity and quality of its Joy, which is 100 Hedons per second, is much greater than that of its Shock, which is 0.1 Algons per second, even though the total *quantity* of energy (or Experience) involved is conserved.

[151] For further discussion of the idea that life has evolved to increase its experience of pleasure see Hameroff (2017) and the work of the philosopher Nathaniel Barrett and his book *Enjoyment as Enriched Experience* (2023).

[152] 'Plural-aspect monism' is a phrase that conveys those views, of which Schrödinger's is an example and dual-aspect monism is another, that take nature to be both one thing and more than one thing depending on the point of view. See Goonatilake (2000) for a discussion of Mach, Schrödinger and Vendantism.

[153] Schrödinger (1958/1992).

directly confronts the problem that so perplexed Charles Sherrington (cited in part I): if mind is a physical process going on in the brain, and physical processes entail interactions between matter and energy, then where in those brain processes do we find mind? Schrödinger applauds Sherrington's "almost brutal" honesty in acknowledging that ignorance on this question is for biology, and indeed science more generally, embarrassing.[154]

Schrödinger's response is to restate his monism, but also to assert that mental states are not dependent for their existence on human brains, especially our brains as they have evolved latterly. Rather, he sees the universe as imbued to its core and through all time with a capacity for mind. All life depends for its existence on of this universal mind and each individual human consciousness is but a fragmentary expression of it. Even though Schrödinger declines to endorse Fechner's view that all matter has a form of mind, this view with its panpsychist implications chimes with the distinction between Experience and conscious Experience that was outlined in part I.

Whatever credence we might give to Schrödinger's views on these matters, and on the role of negentropy in the evolution of living systems, we cannot deny his ambition in seeking to identify deep and universal principles with which to explain great questions in science and philosophy. Perhaps one of the most significant but least commented upon aspects of *What is Life?* is that it is not so much a theory about the physical basis of life, as it is often taken to be. Rather, it constitutes the beginnings of a theory about the physical basis of life *and* mind. The still mysterious nature and origins of these two phenomena are often treated as distinct scientific problems. Similarly, it is

---

[154] Sherrington (1940) called our inability to connect mind and brain an embarrassment for biology, and it remains so today, perhaps more so given the money and intellectual effort expended on the problem since.

often assumed that life must have evolved *before* mind as the existence of the latter must somehow depend on or arise from the former.

As briefly presented and occasionally ill-defined as they are, Schrödinger's proposals in *What is Life?* nevertheless point towards a more unitary vision of nature in which life and mind are aspects of the same physical, chemical and biological processes. Put in terms of the framework outlined here, the tendency of material systems to self-organise into complex structures that absorb, store, mobilise and dissipate energy and so acquire the characteristics of living organisms may be an expression of the very same tendency of systems to self-organise to maximise their production of psychological states of Relief over Distress, on average over time.[155]

On this basis, we can envisage a revision of the widely held assumption that mind supervenes on life or that mental processes depend on physical ones: we may find instead that the causal agency of the intrinsic psychological properties of matter—like those defined herein—exerts an equal or *even greater* influence on the evolution and behaviour of organisms than the traditional extrinsic physical laws and properties. Using the psychophysical framework outlined here, this proposal can be studied theoretically and tested experimentally and may lead to deeper and more general explanations of that evolution and behaviour.

---

[155] One of the reasons I am reluctant to categorise the present framework as panpsychist (in the sense of 'consciousness is everywhere') is that I do not think we should necessarily limit the intrinsic properties of matter to the mental domain only. Although less attention has been paid in this treatise to the nature of life than mind, the framework proposed here is—in spirit at least—as much hylozoic (meaning 'all matter has life') in its implications as it is panpsychist. As Isaac Newton said: "We cannot say that all nature is not alive" (quoted above in part III.2). On the proposed continuity between life and mind see Thompson (2007) and on the related topic of 'biopsychism', Thompson (2022).

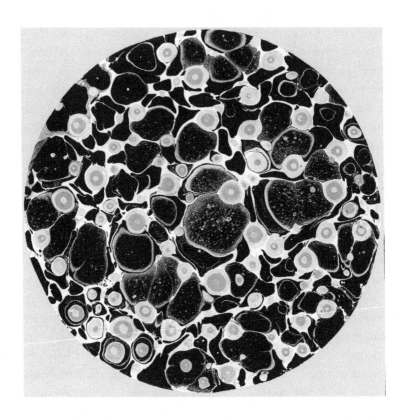

38. *The play of forces is at a standstill, the maximum of entropy attainable for the given system of constraints has been reached (Rudolf Arnheim)*, 2023. Marbled paint on paper, 60 × 60 cm (original in colour). Why do some things look orderly, attractive or beautiful to us while other things look disorderly, ugly or unpleasant? The psychologist and art historian Rudolf Arnheim, whose words provide the title of this artwork, argued that it is the relationship between the play of forces in the objects we perceive and those occurring in the neural processes of our brains that determines our aesthetic responses.

III.8   Rudolf Arnheim's order

In 1854, when he was at the peak of his scientific reputation, Michael Faraday wrote to thank the author Charles Woolnough for a copy of his recently published manual on the art of paper marbling—the first of its kind in the English language.[156] Faraday was familiar with paper marbling as it was widely used in bookbinding, the trade he had apprenticed in before becoming a laboratory assistant to Sir Humphry Davy in 1813. In his letter to Woolnough, he referred to "the very beautiful principles of natural philosophy" involved in this technique, which the book had catalogued and exemplified in detail.[157]

Even though much is left to chance in the behaviour of the (often volatile) media used in the paper marbling process (and we might say, following Schrödinger, that the process has a stochastic tendency towards entropic disorder), it frequently generates patterns that resemble stones, animal markings, eyes, shells, veins and cells (see the image in figure 38). The spontaneity with which these orderly patterns appear suggests there is some common—perhaps deep—organisational principle at work in the physical processes of paper marbling and the formation of certain organic structures.

Many people are attracted, as Faraday was, by this orderliness and naturalistic beauty of marbled papers. What are the "beautiful principles" that Faraday referred to in his letter to Woolnough? How are they expressed in physical processes like marbling and the formation of organic structures, and why do people perceive them as orderly and judge them as beautiful?

---

[156] Woolnough (1853). Some of the material in this section was previously published in Pepperell (2024b).
[157] Faraday (1854).

There have been many attempts over many years in many scientific disciplines and philosophical schools to understand the nature of perceived order and beauty. But very few have tried to explain these phenomena using the principles of physics. One such attempt was made in the early 1970s in a short essay called *Entropy and Art* by Rudolf Arnheim, an art theorist and perceptual psychologist who was trained in the Gestalt school of psychology.[158] This highly influential school, which emerged in Austro-Germany in the late nineteenth and early twentieth centuries, extended Gustav Fechner's psychophysical approach by seeking to ground psychological phenomena in prevailing theories of fundamental physics.

In fact, the founders of the Gestalt school turned to Michael Faraday's work on fields of force when trying to explain the psychophysical nature of perceptual and aesthetic experiences. In doing so they derived two key explanatory principles: the 'psychophysical isomorphism principle', which posits a direct correspondence between psychological states and physical states of the brain, and the 'stability principle', which considers that the tendency of physical systems to move towards maximum stability, or equilibrium, is inherently good, positive or pleasurable. I will consider each of these principles in turn.

In his book *Dynamics in Psychology*, one of the founders of the Gestalt school, Wolfgang Köhler, argued that any neurobiological theory of perception must be a *"field theory"* (his emphasis), like that which evolved from Faraday's experimental work on electromagnetism and was later formalised mathematically by James Clerk Maxwell.[159] As Köhler put it:

---

[158] Arnheim (1971).
[159] Köhler (1940). As a student Köhler had studied physics under Max Planck, who, like Erwin Schrödinger, was major contributor to the development of quantum physics.

…in the theory of perception we are now in precisely the situation in which Faraday found himself when he investigated electrostatic, electromagnetic, and electrodynamic interactions.

According to this theory, the association between "neural functions" and "perceptual facts" (to use Köhler's terminology) is due to psychophysical isomorphism between experience A and brain process *a*, such that for any variation in A there is a corresponding variation in *a*. Consequently, psychological states are identical to physical fields of force at work within the brain, or within the nervous system more widely. This isomorphism principle has a rather remarkable implication: when we perceive an object in the world, the form of the object that we perceive—which is often regarded as purely psychological in nature, even if it might reflect underlying patterns of physical stimulation—is in *itself* a physical system of interacting fields, or what one of Köhler's colleagues, Kurt Koffka, called "psychophysical fields", albeit experienced intrinsically from the perceiver's perspective (and hence invisible to an observer of the fields). In sum, perceptual experiences, like everything else in nature, are manifestations of fields.[160]

Koffka further developed this psychophysical model of perception, which Köhler had originally proposed in the 1920s, by introducing the stability principle. Physics, Koffka noted, teaches us that the world can be considered as a system of interactions between fields of force, where forces are understood as distributed "strains and stresses".[161] The behaviour of objects in the world, including our nervous systems, is governed by laws that depend on the properties and interactions of these fields.

---

[160] Although the relationship between electromagnetic fields, neural activity and mental states was just beginning to be revealed at this time, it is interesting to note the affinities between these Gestalt ideas and the more recent work being done on electromagnetic field theories of consciousness, as discussed in part III.5.
[161] Koffka (1935).

Köhler and Koffka state that the organisation of objects in the world is bound to take certain forms because all processes must follow the same physical laws that compel systems towards stability and the minimisation of energy in the relevant fields; in any system where the forces exerted in the fields are imbalanced the system will act to achieve the greatest balance most quickly, thus reducing the strains and stresses acting on it to the greatest extent possible. As expressed by Köhler in his law of dynamic direction in 1920:

> In all processes which terminate in time dependent states the distribution shifts towards a minimum of energy.[162]

To illustrate this he gives the example of a soap bubble, which is the simplest, most stable and most orderly form given all the opposing forces that act on its matter and the quantity of energy supplied to do the work that produces it.[163] In one sense, the stability principle is related to the principle by which dissipative structures emerge, as outlined in the previous section. Both principles describe the tendency of systems to self-organise into orderly patterns when suitable kinds and quantities of energy are supplied to systems having sufficient opportunity for self-reconfiguration. The important difference, however, is that the Gestalt theorists had in mind systems that tended towards (approximately) 'static' thermodynamic equilibrium. Dissipative structures, by contrast, are generally in a dynamic and far-from-equilibrium condition, which is sustained by inward energy flows. But the scientific concept of dynamic or far-from-equilibrium steady-states that are

---

[162] Quoted from Koffka (1935). By "time dependent states" Koffka means states of (approximately) 'static' or thermodynamic equilibrium.

[163] The same stabilising process is evident in the artwork reproduced in figure 37, in which combining immiscible substances caused them to spontaneously form into numerous spheres as the system transformed from a higher to a lower entropic state.

driven by continuous influx, storage and mobilisation of free energy, as discussed above in part III.7, was not developed until later in the twentieth century.

Another founder of the Gestalt school, Max Wertheimer, introduced the "law of Prägnanz" to account for the operation of energy minimisation and equilibration principles in the domain of psychophysics. This law mandated that the "psychological organisation will always be as 'good' as the prevailing conditions allow".[164] In formulating this law, Wertheimer was building on the psychophysical isomorphism and stability principles propounded by his colleagues to establish a direct relationship between:

1. the organisation and behaviour of material objects in the world due to interactions between fields of force and their tendency towards order or stability,

2. the organisation of the fields of force acting in the matter of the sensory systems of those who perceive the material objects and how those forces affect the matter concerned, and

3. the specific organisation and hedonic valence that characterises perceptually experiencing the objects.[165]

---

[164] Cited in Koffka (1935). For a recent review on the law of Prägnanz—its history, development and relation to visual perception—see Van Geert and Wagemans (2024).

[165] It is worth noting that despite Wertheimer's stress on the 'goodness' of form it was also implicit in the law of Prägnanz that systems being displaced from stability result in a 'bad' perceptual organisation and a concomitant negative hedonic experience. It is also worth noting the conceptual similarity between these three points and the three Conjectures summarised in part II.3.

The psychophysical principles driving this threefold relationship were subsumed into the "fundamental law of aesthetics" by another proponent of Gestalt psychology, Hans Eysenck, according to whom:

> The pleasure derived from a percept as such is directly proportional to the decrease of energy capable of doing work in the total nervous system, as compared with the original state of the whole system.[166]

By applying these Gestalt principles of good (and bad) form to the operation of the nervous system, Eysenck, like his colleagues, was advancing a neurobiological theory of aesthetic value and hedonic valence derived from the fundamental physical laws by which fields and forces operate in the world and in nervous systems.

Rudolf Arnheim further extended this school of thought through the ideas he published *Entropy and Art*. In doing so he drew on many of the nineteenth and early twentieth century scientists mentioned in this treatise—including Johannes Zöllner, Sigmund Freud and Gustav Fechner, as well as the Gestalt school figures already mentioned—to formulate a theory of aesthetics grounded in the psychophysical economy of energy minimisation.

He begins the book—which is subtitled '*An Essay on Disorder and Order*'—by noting that "order is a necessary condition for making a structure function" and "a prerequisite of survival". For Arnheim, it is only by being ordered that anything can exist in the world and can be known to exist through the orderly functioning of our minds. But Arnheim also follows Schrödinger in recognising that order is the

---

[166] Eysenck (1942). Although he did not state it, the inverse of this law—that *displeasure* is directly proportional to the *increase* of energy—would be equally true. For a recent critical discussion of these Gestalt theories and their application by Arnheim to processes of aesthetic appreciation, see Poulaki (2022).

complement of disorder, and that two seemingly conflicting tendencies are perpetually at work in the cosmos: the tendency towards disintegration and decay, which is often thought to be mandated by the Second Law of Thermodynamics, and the tendency towards integration and organisation, which we observe in the spontaneously forming structures that exist throughout nature and which we ourselves exemplify and embody.

To resolve the apparent conflict, Arnheim invokes a further law of Gestalt theory, as noted above: Köhler's law of dynamic direction. It is through obeying this law, according to Arnheim, that what he calls the "structural order" emerges in nature in a way that is closely related to the emergence of dissipative structures, as discussed earlier in part III.7. In many kinds of systems, the effect of the tendency towards energy minimisation and thermodynamic equilibrium is to produce entropically more probable states of disintegration, decay or disorder. But in other kinds of systems the effect is the opposite: if enough of the right kind of energy is flowing through a system having enough opportunity for self-reconfiguration, then the optimal way for a system to minimise its energy and equilibrate is to organise itself into a more orderly and less probable structure.

Arnheim then applies these laws and principles to the processes involved in the creation and appreciation of art. For Arnheim, one of the ways in which pleasure is gained through art is by perceiving physical forms that optimally reduce the competing forces that act, or appear to act, upon them or within them. This leads to a corresponding production of and reduction of tension (strain and stress) in the nervous system and so in the mind of the beholder, as per the Gestalt psychophysical isomorphism and stability principles.

Even though paintings and sculptures (and visual artworks of other kinds) are usually 'frozen' or static arrangements of matter, such as the paper marbling piece shown in figure 38 or Faraday's drawings of iron filings shown in figure 34, Arnheim asks us to consider them not as inert arrangements of matter—that is, as fixed patches of pigment on paper or iron filings stuck in wax—but as potentially dynamic structures composed through the "antagonistic play of forces" in fields which are tending towards the simplest and most stable configuration they can adopt under the circumstances. The dynamic processes of energy flow and entropy production that lead to a finished work of art, and which are embodied in its final state, are then isomorphically instantiated in our minds when we behold it. Concerning the sense of beauty evoked by the sculpture of the Madonna of Würzburg (reproduced here in figure 39), and the way the sculptor has invested the figure's posture with a sense of dynamism, weight and poise, he says:

> Each element has its appropriate form in relation to all others, thus establishing a definitive order, in which all component forces hold one another in such a way that none of them can press for any change of the interrelation. The play of forces is at a standstill, the maximum of entropy attainable for the given system of constraints has been reached. Although the tension invested in the work is at a high absolute level, it is reduced to the lowest level the constraints will let it assume.[167]

According to Arnheim, it is because the sculptor has so exquisitely managed the balance of competing forces represented in the sculpture that the competing forces acting in the fields in our brains, and so in our minds, are likewise exquisitely balanced as we perceive it. The parallel experience we have is intrinsically pleasurable because it affords the forces acting in our minds an opportunity for tension creation and

---

[167] Arnheim (1971).

reduction as the equilibrated balance is reached. It is to this predominantly positive hedonic valence, and the object that evokes it, that we attribute the value of 'beauty'.

The reason that processes of tension reduction and equilibration lead to enhanced orderliness and dynamism, rather than entropically more probable disorderliness and inertia, is because of the opportunities for dynamic reconfiguration and increasing organisational complexity afforded to our perceptual systems (which Koffka understood in terms of psychophysical fields) as they are formed in response to the work.

We could apply this analysis to account for the beauty of the organic patterns we see in marbled papers. The technique of paper marbling exploits several principles of classical mechanics and thermodynamics. The method can be viewed as creating a dynamic material system in which portions of matter (drops of coloured media) containing certain quantities of thermal and chemical energy have further kinetic energy transferred to them during the process of application, which is then dissipated as the matter contacts the surface of the bath.

Forces acting on the media during application are directed downwards due to gravity and outwards from the centre of application due to the bath's surface tension, causing the media to expand into a thin disc. This disc reaches its lowest-energy state and largest size when its cohesive forces prevent further expansion. More energy and matter are added to the system as new drops of media are applied, which do mechanical work on existing discs and move their boundaries. Forces of repulsion, adhesion and cohesion in different portions of media act on each other such that the pattern on the surface of the bath becomes a dynamic system that seeks to accommodate all its opposing forces.

Once the addition of new media ceases, the sum effect of all the forces that have been exerted in the system produces an equilibrated pattern that constitutes the state of the system with the lowest free energy. Finally, a print is taken from the pattern by carefully laying paper on the surface of the bath such that the floating media transfer to the paper, permanently capturing the equilibrium of the system in a design.

The apparent orderliness of the final marbled design is determined to some extent by the skill and experience of the marbler. But as any novice marbler knows, the marbling media have a propensity to disintegrate into a formless mess given the highly stochastic tendencies of the matter and energy in the system. When the media are suitably controlled, however, the chances increase greatly for stable and often beautiful forms to emerge spontaneously.

We can think of the paper marbling process and the images it produces as exemplifying the psychophysical principles of nature, art and aesthetics discussed by Arnheim. These principles rest on the tension and balance between the conflicting cosmic tendencies towards emergent order and stochastic disorder. The forms generated by the optimal balancing of these two tendencies in nature, in minds and in art due to the actions of forces on matter, may express the "beautiful principles of natural philosophy" that Faraday referred to.

Using the psychophysical explanatory framework outlined in this treatise, which is consistent with the Gestalt approach adopted by Arnheim in his essay, we can begin to understand how complex and orderly forms in physical systems arise, how perceiving them affects the organisation of our nervous systems, and how we respond to these perceptions psychologically—that is, aesthetically, hedonically, emotionally and perhaps even spiritually—from our intrinsic perspectives.

39. A reproduction of the photograph of the Madonna of Würzburg, or the Gotische Madonna, of 1420 as discussed by Rudolf Arnheim in his book *Entropy and Art*. The image is taken from a postcard in which the photograph is credited to Eberhard Zwicker.

Coda 3

As we have just seen in part III.8, some of the foundations of the explanatory framework outlined in this treatise were laid by the Gestalt theorists of the early twentieth century, as well as by figures like Gustav Fechner and Ernst Mach before them. Rudolf Arnheim built on this groundwork in *Entropy and Art* and extended it further into the realms of art and aesthetics. I hope I have shown that the framework outlined in this treatise—which by necessity integrates knowledge from the arts, sciences and humanities—could be used to generate falsifiable hypotheses and predictions that can be experimentally tested using current methods and tools. The results of these experiments may contribute to a greater understanding of the problem set out at the start, which is how matter gets its mind, and why.

At stated in the preface, the aim of this treatise has been to provide a scientific framework within which to address the longstanding problem of how mind relates to matter and to better explain how conscious experiences come to exist in the world. For what it's worth, my own hunch is that the density, subtlety and intricacy of the psychophysical processes that enable consciousness to occur may be of such a high order that, although entailing nothing more than interactions between matter and energy, they may lie beyond our current understanding.[168] But I hope that by addressing the problem in the way proposed here— that is, by taking the conceptual framework seriously and testing it experimentally—and by systematically combining knowledge from diverse fields, we can at least, and at last, begin to solve the problem.

---

[168] As suggested elsewhere in this treatise, I suspect we will ultimately find the explanation for consciousness among the energy fields that resonate at different powers and frequencies and interact in exquisitely coordinated ways with the matter of nervous systems—and other bodily tissues—over a vast range of spatiotemporal scales.

Bibliography

Adrian, E. (1928). *The basis of sensation*. W. W. Norton & Co.

Andrade-Talavera, Y., Fisahn, A., & Rodríguez-Moreno, A. (2023). Timing to be precise? An overview of spike timing-dependent plasticity, brain rhythmicity, and glial cells interplay within neuronal circuits. *Molecular Psychiatry*, *28*(6), 2177–2188.

Arnheim, R. (1971). *Entropy and art*. University of California Press.

Arnheim, R. (1985). The other Gustav Theodor Fechner. In S. Koch & D. E. Leary (Eds.), *A century of psychology as science* (pp. 856–865). American Psychological Association.

Assis, A. K. T., & Karam, R. (2018). The free fall of an apple: Conceptual subtleties and implications for physics teaching. *European Journal of Physics*, *39*, Article 035003.

Atasoy, S., Donnelly, I., & Pearson, J. (2016). Human brain networks function in connectome-specific harmonic waves. *Nature Communications*, *7*, Article 10340.

Atmanspacher, H. (2012). Dual-aspect monism à la Pauli and Jung. *Journal of Consciousness Studies 19*(9–10), 96–120.

Baars, B., Franklin, S., & Ramsoy, T. (2013). Global workspace dynamics: Cortical "binding and propagation" enables conscious contents. *Frontiers in Psychology*, *4*, Article 200.

Ball, P. (2023). *Organisms as agents of evolution*. John Templeton Foundation.

Barrett, N. (2023). *Enjoyment as enriched experience*. Palgrave Macmillan Cham.

Bartlett, G. (2022). Does integrated information theory make testable predictions about the role of silent neurons in consciousness? *Neuroscience of Consciousness*, *2022*(1), Article niac015.

Bazzigaluppi, P., Amini, A. E., Weisspapier, I., Stefanovic, B., & Carlen, P. (2017). Hungry neurons: Metabolic insights on seizure dynamics. *International Journal of Molecular Sciences*, *18*(11), Article 2269.

Bentham, J. (1789). *Introduction to the principles of morals and legislation* (J. Bowring, Ed.). W. Pickering.

Bigelow, J., Ellis, B., & Pargetter, R. (1988). Forces. *Philosophy of Science*, *55*(4), 614–630.

Boswell, J. (1791). *The life of Samuel Johnson*. Charles Dilly.

Cabral, J., Fernandes, F. F., & Shemesh, N. (2023). Intrinsic macroscale oscillatory modes driving long range functional connectivity in female rat brains detected by ultrafast fMRI. *Nature Communications, 14*, Article 375.

Casali, A. G., Gosseries, O., Rosanova, M., Boly, M., Sarasso, S., Casali, K. R., Casarotto, S., Bruno, M.-A., Laureys, S., Tononi, G., & Massimini, M. (2013). A theoretically based index of consciousness independent of sensory processing and behavior. *Science Translational Medicine, 5*(198), Article 198ra105.

Chalmers, D. (2007). Naturalistic Dualism. In M. Velmans & S. Schneider, (Eds.), *The Blackwell Companion to Consciousness* (pp. 359-368). Blackwell Publishing.

Clausberg, K. (2007). Feeling embodied in vision: The imagery of self-perception without mirrors. In J. Krois, M. Rodengren, A. Steidele, & D. Westerkamp (Eds.), *Embodiment and cognition* (pp. 77–103). John Benjamins.

Clausius, R. (1867). *The mechanical theory of heat – with its applications to the steam engine and to physical properties of bodies*. John van Voorst.

Clifford, W. K. (1878). On the nature of things-in-themselves. *Mind, 3*(9), 57–67.

Coopersmith, J. (2010). *Energy, the subtle concept: The discovery of Feynman's blocks from Leibniz to Einstein*. Oxford University Press.

Crane, T. (1995). The mental causation debate. *Proceedings of the Aristotelian Society, 69*(Suppl.), 211–236.

Déli, É., Peters, J. F., & Kisvárday, Z. (2022). How the brain becomes the mind: Can thermodynamics explain the emergence and nature of emotions? *Entropy (Basel), 24*(10), Article 1498.

Descartes, R. (1972). *Treatise of man*. Cambridge, Mass.: Harvard University Press. (Thomas Steele Hall, Tr.). (Original work published 1662).

Descartes, R. (1983). *Principles of Philosophy*. Dordrecht: D. Reidel Publishing Company. (V. R. Miller and R. P. Miller, Tr.). (Original work published 1644).

Descartes, R. (1989). *Correspondance avec Elisabeth* (J.-M. Beyssade & M. Beyssade, Eds.). Garnier-Flammarion.

Descartes, R. (2006). *A discourse on the method of correctly conducting one's reason and seeking truth in the sciences* (I. Maclean, Trans.). Clarendon Press. (Original work published 1637).

Dinuzzo M., & Nedergaard M. (2017). Brain energetics during the sleep–wake cycle. *Current Opinion in Neurobiology*, *47*, 65–72.

Dobelle, W. H., & Mladejovsky, M. G. (1974). Phosphenes produced by electrical stimulation of human occipital cortex, and their application to the development of a prosthesis for the blind. *The Journal of Physiology*, *243*(2), 553–576.

du Châtelet, É. (1740). *Institutions physique [Foundations of physics]*. Aux Dépens de la Compagnie.

Dyson, F. (1953). Field theory. *Scientific American*, *188*(4), 57–65.

Eddington, A. (1928). *The nature of the physical world*. Cambridge University Press.

Edelman G.M., & Gally, J.A. (2013). Reentry: A key mechanism for integration of brain function. *Frontiers in Integrative Neuroscience*, *7*, Article 63.

Einstein, A. (2002). Fundamental ideas and methods of the theory of relativity, presented in their development (A. Engel, Trans.). In *The collected papers of Albert Einstein, volume 7: The Berlin years: Writings, 1918–1921* (pp. 113–150). Princeton University Press. (Original work published 1920).

Einstein, A. (2012). Response to Ernest Bovet's question to Paul Langevin (A. Engel, Trans.). In *The collected papers of Albert Einstein, volume 13: The Berlin years: Writings & correspondence, January 1922 – March 1923* (p. 181). Princeton University Press. (Original work published 1922).

Endres, R. G. (2017). Entropy production selects nonequilibrium states in multistable systems. *Scientific Reports*, *7*, Article 14437.

England, J. L. (2015). Dissipative adaptation in driven self-assembly. *Nature Nanotechnology*, *10*(11), 919–923.

England, J. (2020). *Every life is on fire. How thermodynamics explains the origins of living things*. Basic Books.

Eysenck, H. (1942). The experimental study of the 'good Gestalt'—a new approach. *Psychological Review*, *49*(4), 344–364.

Faraday, M. (1832). Experimental researches in electricity. *Philosophical Transactions of the Royal Society of London*, *122*, 125–162.

Faraday, M. (1852). LVIII. On the physical character of lines of magnetic force. *The London, Edinburgh, and Dublin Philosophical Magazine and Journal of Science (Fourth Series), 3*(20), 401–428.

Faraday, M. (1999). Faraday to Charles W. Woolnough, 2 January 1854. *Epsilon: The Michael Faraday Collection*, Article Faraday2773. https://epsilon.ac.uk/view/faraday/letters/Faraday2773 (Original work published 1854).

Faraday, M. (1855). *Experimental researches in electricity, Vol. 3*. Richard and John E. Taylor.

Faraday, M. (1859). *A course of six lectures on the various forces of matter and their relations to each other* (3rd ed.). Richard Griffin and Company.

Fechner, G. (1904). *The little book of life after death*. (M. Wadsworth, Trans.). Little, Brown & Company. (Original work published 1836).

Fechner, G. (1966). *Elements of psychophysics. Vol. 1* (H. R. Adler, D. H. Howes, & E. G. Boring, Trans.). Holt, Rinehart and Winston. (Original work published 1860).

Feynman, R. (1964). *The Feynman lectures on physics*. Basic Books.

Francken, J. C., Beerendonk, L., Molenaar, D., Fahrenfort, J. J., Kiverstein, J. D., Seth, A. K., & van Gaal, S. (2022). An academic survey on theoretical foundations, common assumptions and the current state of consciousness science. *Neuroscience of Consciousness, 2022*(1), Article niac011.

Freud, S. (1913). *The interpretation of dreams* (A. A. Brill, Trans.). The Macmillan Company. (Original work published 1900).

Freud, S. (1950). *Project for a scientific psychology: Volume 1 of standard edition of the complete psychological works of Sigmund Freud* (J. Strachey, Trans.). Hogarth Press. (Original work published 1895).

Friston, K. (2010). The free-energy principle: A unified brain theory? *Nature Reviews Neuroscience, 11*, 127–138.

Gilad, A. (2024). Wide-field imaging in behaving mice as a tool to study cognitive function. *Neurophotonics, 11*(3), Article 033404.

Gitterman, M., & Halpern, V. (2004). *Phase transitions: A brief account with modern applications*. World Scientific.

Goff, P., & Moran, A. (2022). *Is consciousness everywhere? Essays on panpsychism*. Imprint Academic.

Goonatilake, S. (2000). Many paths to enlightenment. *Nature, 405*(6785), 399.

Haeckel, E. (1901). *The riddle of the universe at the close of the nineteenth century* (2nd ed.) (J. McCabe, Trans.). Watts & Co.

Guillery R. (2005). Observations of synaptic structures: origins of the neuron doctrine and its current status. *Philosophical transactions of the Royal Society of London. Series B, Biological sciences, 360*(1458), 1281–1307.

Hagengruber, R. (Ed.). (2012). *Émilie du Châtelet between Leibniz and Newton.* Springer Dordrecht.

Hales, C. G. (2014). The origins of the brain's endogenous electromagnetic field and its relationship to provision of consciousness. *Journal of Integrative Neuroscience, 13*(2), 313–361.

Hameroff, S. (2017). The quantum origin of life: How the brain evolved to feel good. In M. Tibayrenc & F. J. Ayala (Eds.), *On human nature: Biology, psychology, ethics, politics, and religion* (pp. 333–353). Academic Press.

Hameroff, S., & Penrose, R. (2014). Consciousness in the universe: A review of the 'Orch OR' theory. *Physics of Life Reviews, 11*(1), 39–78.

Hecht, E. (2019). Understanding energy as a subtle concept: A model for teaching and learning energy. *American Journal of Physics, 87*, 495–503.

Heffern, E. F. W., Huelskamp, H., Bahar, S., & Inglis, R. F. (2021). Phase transitions in biology: From bird flocks to population dynamics. *Proceedings of the Royal Society B: Biological Sciences, 288*(1961), Article 20211111.

Heidelberger, M. (1994). The unity of nature and mind: Gustav Theodor Fechner's non-reductive materialism. In S. Poggi & M. Bossi (Eds.), *Romanticism in science* (pp. 215–236). Springer Dordrecht.

Heidelberger, M. (2004). *Nature from within: Gustav Theodor Fechner and his psychophysical worldview.* University of Pittsburgh Press.

Heisenberg, W. (1958). *Physics and philosophy: The revolution in modern science* (R. N. Anshen, Ed.). Harper & Brothers Publishers.

Herculano-Houzel, S. (2012). The remarkable, yet not extraordinary, human brain as a scaled-up primate brain and its associated cost. *Proceedings of the National Academy of Sciences of the United States of America, 109*(Suppl. 1), 10661–10668.

Ho, M.-W. (1994). What is (Schrödinger's) negentropy? *Modern Trends in Bio-ThermoKinetics, 3*, 50–61.

Ho, M.-W. (2008). *The rainbow and the worm: The physics of organisms* (3rd ed.). World Scientific.

Hudetz, A. G., & Mashour, G. A. (2016). Disconnecting consciousness: Is there a common anesthetic end point? *Anesthesia & Analgesia, 123(5),* 1228–1240.

Hunt, T., & Schooler, J. W. (2019). The easy part of the hard problem: A resonance theory of consciousness. *Frontiers in Human Neuroscience, 13,* Article 378.

Hunt, T., Jones, M., McFadden, J., Delorme, A., Hales, C. G., Ericson, M., & Schooler, J. (2024). Editorial: Electromagnetic field theories of consciousness: Opportunities and obstacles. *Frontiers in Human Neuroscience, 17,* Article 1342634.

Iliffe, R. (2017). *Priest of nature: The religious worlds of Isaac Newton.* Oxford University Press.

Iltis, C. (1971). Leibniz and the vis viva controversy. *Isis, 62*(1), 21–35.

Iwanski, M. K., & Kapitein, L. C. (2023). Cellular cartography: Towards an atlas of the neuronal microtubule cytoskeleton. *Frontiers in Cell and Developmental Biology, 11,* Article 1052245.

James, W. (1890). *The principles of psychology.* Henry Holt.

Jenny, H. (2007). *Cymatics: A study of wave phenomena and vibration.* Macromedia Press.

Jones, M. W., & Hunt, T. (2023). Electromagnetic-field theories of qualia: Can they improve upon standard neuroscience? *Frontiers in Psychology, 14,* Article 1015967.

Jones, M. (2013). Electromagnetic-field theories of mind. *Journal of Consciousness Studies, 20*(11–12), 124–149.

Jorati, J. (2015). Leibniz on causation – part 1. *Philosophy Compass, 10*(6), 389–397.

Kachman, T., Owen, J. A., & England, J. L. (2017). Self-organized resonance during search of a diverse chemical space. *Physical Review Letters, 119,* Article 038001.

Köhler, W. (1940). *Dynamics in psychology.* Liveright Publishing Corporation.

Kondepudi, D., & Prigogine, I. (2015) *Modern thermodynamics: From heat engines to dissipative structures.* John Wiley & Sons.

Kondepudi, D. K., De Bari, B., & Dixon, J. A. (2020). Dissipative structures, organisms and evolution. *Entropy (Basel)*, *22*(11), Article 1305.

Koffka, K. (1935). *Principles of Gestalt psychology*. London: Routledge & Kegan Paul Ltd.

Kuhn, R. L. (2024). A landscape of consciousness: Toward a taxonomy of explanations and implications. *Progress in Biophysics and Molecular Biology*, *190*, 28–169.

Lambert, F. (2002). Entropy is simple, qualitatively. *Journal of Chemical Education*, *79*(10), 1241-1246.

Lane, N. (2022). *Transformer: The deep chemistry of life and death*. Profile Books.

Lau, H., & Rosenthal, D. (2011). Empirical support for higher-order theories of conscious awareness. *Trends in Cognitive Sciences*, *15*(8), 365–373.

Leibniz, G. (1686). *Discours de métaphysique [Discourse on metaphysics]*.

Leibniz, G. (1960). *Die philosophischen schriften von Gottfried Wilhelm Leibniz, Vols. 1–7* (C. I. Gerhardt, Ed.). Olms Verlagsbuchhandlung. (Original works published 1875–1890 by Weidmannsche Buchhandlung).

Levere, T. (1968). Faraday, matter and natural theology: Reflections on an unpublished manuscript. *The British Journal for the History of Science*, *4*(2), 95–107.

Levin, M. (2019). The computational boundary of a 'self': Developmental bioelectricity drives multicellularity and scale-free cognition. *Frontiers in Psychology*, *10*, Article 2688.

Levin, M. (2021). Bioelectric signaling: Reprogrammable circuits underlying embryogenesis, regeneration, and cancer. *Cell*, *184*(8), 1971–1989.

Li, T., Zheng, Y., Wang, Z., Zhu, D. C., Ren, J., Liu, T., & Friston, K. (2022). Brain information processing capacity modeling. *Scientific Reports*, *12*, Article 2174.

Locke, J. (1690). *An essay concerning human understanding*. Thomas Basset.

Lodge, P. (2014). Leibniz's mill argument against mechanical materialism revisited. *Ergo: An Open Access Journal of Philosophy*, *1*(3), 79–99.

Mach, E. (1914). *Analysis of the sensations, and the relation of the physical to the psychical* (C. M. Williams, Trans.). The Open Court Publishing Company. (Original work published 1897).

Mach, E. (1960). *The science of mechanics* (T. J. McCormack, Trans.). The Open Court Publishing Company. (Original work published 1883).

Magistretti, P. J., & Allaman, I. (2013). Brain energy metabolism. In D. W. Pfaff (Ed.), *Neuroscience in the 21st century* (pp. 1591–1620). Springer New York, NY.

Mashour, G. A., Roelfsema, P., Changeux, J.-P., & Dehaene, S. (2020). Conscious processing and the global neuronal workspace hypothesis. *Neuron*, *105*(5), 776–798.

Massimini, M., Sarasso, S., Casarotto, S., & Rosanova, M. (2022). Measures of differentiation and integration: One step closer to consciousness. *The Behavioral and Brain Sciences*, *45*, Article e54.

Maxwell, J. C. (1877). *Matter and motion*. Society for Promoting Christian Knowledge.

McFadden, J. (2020). Integrating information in the brain's EM field: The cemi field theory of consciousness. *Neuroscience of Consciousness*, *2020*(1), Article niaa016.

McFadden, J. (2023). Consciousness: Matter or EMF? *Frontiers in Human Neuroscience*, *16*, Article 1024934.

McGuire, J. E. (1968). Force, active principles, and Newton's invisible realm. *Ambix*, *15*(3), 154–208.

McMullin, E. (2002). The origins of the field concept in physics. *Physics in Perspective*, *4*, 13–39.

Mizukami, H., Kakigi, R., & Nakata, H. (2019). Effects of stimulus intensity and auditory white noise on human somatosensory cognitive processing: A study using event-related potentials. *Experimental Brain Research*, *237*, 521–530.

Nagel, T. (1974). What is it like to be a bat? *The Philosophical Review*, *83*(4), 435–450.

Nägeli, C. (1877). The limits of natural knowledge II. *Nature*, *15*, 549.

Newton, I., Cohen, B., & Whitman, A. (1999). *The Principia: Mathematical principles of natural philosophy*. University of California Press.

Owen, A., Coleman, M., Boly, M., Davis, M., Laureys, S., & Pickard, J. (2006). Detecting awareness in the vegetative state. *Science*, *313*(5792), 1402.

Papineau, D. (1977). The vis viva controversy: Do meanings matter? *Studies in History and Philosophy of Science Part A*, *8*(2), 111–142.

Park, W., Kim, S. P., & Eid, M. (2021). Neural coding of vibration intensity. *Frontiers in Neuroscience*, *15*, Article 682113.

Pepperell, R. (2015). Egocentric perspective: Depicting the body from its own point of view. *Leonardo*, *48*(5), 424–429

Pepperell, R. (2018). Consciousness as a physical process caused by the organization of energy in the brain. *Frontiers in Psychology*, *9*, Article 2091.

Pepperell, R. (2020). *Vision as an energy-driven process*. arXiv. https://doi.org/10.48550/arXiv.2008.00754

Pepperell, R. (2024a). Being alive to the world: An artist's perspective on predictive processing. *Philosophical Transactions of the Royal Society B: Biological Sciences*, *379*(1895), Article 20220429.

Pepperell, R. (2024b). 'The very beautiful principles of natural philosophy': Michael Faraday, paper marbling and the physics of natural forms. *LASER Journal*, *2*(1), Article 2.

Piatkevich, K. D., Bensussen, S., Tseng, H. A., Shroff, S. N., Lopez-Huerta, V. G., Park, D., Jung, E. E., Shemesh, O. A., Straub, C., Gritton, H. J., Romano, M. F., Costa, E., Sabatini, B. L., Fu, Z., Boyden, E. S., & Han, X. (2019). Population imaging of neural activity in awake behaving mice. *Nature*, *574*, 413–417.

Pockett, S. (2012) The electromagnetic field theory of consciousness: A testable hypothesis about the characteristics of conscious as opposed to nonconscious fields. *Journal of Consciousness Studies*, *19*(11–12), 191–223.

Poulaki, M. (2022). A Gestalt theory for 'disorder': From Arnheim's ordered chaos to Brambilla's entropic art. *Cinéma & Cie. Film and Media Studies Journal*, *22*(38), 55–67.

Preston, J. (2021). Phenomenalism, or neutral monism, in Mach's *Analysis of Sensations?* In J. Preston (Ed.), *Interpreting Mach: Critical essays* (pp. 235–257). Cambridge University Press.

Reber, A. (2018). *The first minds: Caterpillars, karyotes, and consciousness*. Oxford University Press.

Robb, D., Heil, J., & Gibb, S. (2023) Mental causation. In E. N. Zalta & U. Nodelman (Eds.), *The Stanford encyclopedia of philosophy (Spring 2023 edition)*. Stanford University. https://plato.stanford.edu/archives/spr2023/entries/mental-causation/

Roberts, T. (2015). *Einstein's intuition: Visualizing nature in eleven dimensions*. Quantum Space Theory Institute.

Rocca, M. D. (2021). Spinoza's metaphysical psychology. In D. Garrett (Ed.), *The Cambridge companion to Spinoza* (2nd ed.) (pp. 234–281). Cambridge University Press.

Röhl, M., & Uppenkamp, S. (2012). Neural coding of sound intensity and loudness in the human auditory system. *Journal of the Association for Research in Otolaryngology, 13*(3), 369–379.

's Gravesande, W. (1722). Essai d'une nouvelle theorie sur le choc des corps fondée sur l'expérience [Essay on a new theory of the collision of bodies based on experience]. *Journal Littéraire de la Haye, 12*, 1–54.

Sarasso, S., Casali, A., Casarotto, S., Rosanova, M., Sinigaglia, C., & Massimini, M. (2021). Consciousness and complexity: A consilience of evidence. *Neuroscience of Consciousness, 2021*(2), Article niab023.

Schneider, E. D., & Kay, J. J. (1989). Nature abhors a gradient. In P. Ledington (Ed.), *Proceedings of the 33rd Annual Meeting of the International Society for the System Sciences* (Vol. 3, pp. 19–23). International Society for the Systems Sciences.

Schneider, E., & Sagan, D. (2005). *Into the cool: Energy flow, thermodynamics, and life*. University of Chicago Press.

Schrödinger, E. (1944). *What is life? The physical aspect of the living cell*. Cambridge University Press.

Schrödinger, E. (1992). *Mind and matter*. In *What is life? With mind and matter and autobiographical sketches* (pp. 91–116). Cambridge University Press. (Reprinted from *Mind and matter* by E. Schrödinger, 1958, Cambridge University Press).

Scott, D. (1997). Leibniz and the two clocks. *Journal of the History of Ideas, 58*(3), 445–463.

Seager, W. (2016). Panpsychist infusion. In G. Bruntup & L. Jaskolla (Eds.), *Panpsychism: Contemporary perspectives*. Oxford University Press.

Seager, W. (2022). Panpsychism and energy conservation. *Mind and Matter,* *20*(1), 17–34.

Seth, A. (2021). *Being you: A new science of consciousness.* Faber & Faber.

Seth, A. K., & Bayne, T. (2022). Theories of consciousness. *Nature Reviews Neuroscience, 23*(7), 439–452.

Sherrington, C. (1940). *Man on his nature.* Cambridge University Press.

Skrbina, D. (2007). *Panpsychism in the West.* MIT Press.

Solms, M. (2021). *The hidden spring: A journey to the source of consciousness.* Profile Books.

Spinoza, B. (1954). *Ethics and on the improvement of the understanding* (J. Gutmann, Ed.). Hafner Publishing Company (Original work published 1677).

Staley, R. (2021). Sensory studies, or when physics was *psychophysics*: Ernst Mach and physics between physiology and psychology, 1860–71. *History of Science, 59*(1), 93–118.

Strassler, M. (2024). *Waves in an impossible sea: How everyday life emerges from the cosmic ocean.* Basic Books.

Stukeley, W. (2004, September). *Memoirs of Sir Isaac Newton's life.* The Newton Project. https://www.newtonproject.ox.ac.uk/view/texts/normalized/OTHE00001 (Original work published 1752).

Swenson, R. (2023). A grand unified theory for the unification of physics, life, information and cognition (mind). *Philosophical Transactions of the Royal Society A: Mathematical, Physical and Engineering Sciences, 381*(2252), Article 20220277.

Thompson, E. (2007). *Mind in Life: Biology, Phenomenology, and the Sciences of Mind.* Harvard University Press.

Thompson, E. (2022). Could All Life Be Sentient? *Journal of Consciousness Studies, 29(3).* 229–265.

Toker, D., Pappas, I., Lendner, J. D., Frohlich, J., Mateos, D. M., Muthukumaraswamy, S., Carhart-Harris, R., Paff, M., Vespa, P. M., Monti, M. M., Sommer, F. T., Knight, R. T., & D'Esposito, M. (2022). Consciousness is supported by near-critical slow cortical electrodynamics. *Proceedings of the National Academy of Sciences of the United States of America, 119*(7), Article e2024455119.

Tononi, G., & Edelman, G. M. (1998). Consciousness and complexity. *Science, 282*(5395), 1846–1851.

Tononi, G., Boly, M., Massimini, M., & Koch, C. (2016). Integrated information theory: From consciousness to its physical substrate. *Nature Reviews Neuroscience, 17*, 450–461.

Van Geert, E., & Wagemans, J. (2024). Prägnanz in visual perception. *Psychonomic Bulletin & Review, 31*(2), 541–567.

Warren, J. W. (1982). The nature of energy. *European Journal of Science Education, 4*(3), 295–297.

Westfall, R. (1994). *The life of Isaac Newton.* Cambridge University Press.

Whitehead, A. N. (1929). *Process and reality: An essay in cosmology.* The Free Press.

Williams, L. P. (1960). Michael Faraday and the evolution of the concept of the electric and magnetic field. *Nature, 187*, 7030–7033.

Woolnough, C. W. (1853). *The art of marbling, as applied to book-edges and paper.* Alexander Heylin.

Yokoyama, Y., Yamada, Y., Kosugi, K., Yamada, M., Narita, K., Nakahara, T., Fujiwara, H., Toda, M., & Jinzaki, M. (2021). Effect of gravity on brain structure as indicated on upright computed tomography. *Scientific Reports, 11*, Article 392.

Zöllner, J. (1872). *Über die natur der cometen.* Verlag von Wilhelm Engelmann.

Index of names

(For cited works with more than two authors, the first author is indexed)

ABOUT THE AUTHOR

Robert Pepperell PhD was born in London in 1963 and studied at the Slade School of Art, University College London before embarking on a varied career as artist, VJ, university professor, scientist and inventor. Throughout this career he has tackled deep questions about mind and nature by combining art, philosophy and science. This has led him to author several books and publish articles in many fields, including in philosophy, neuroscience, consciousness studies, psychology, computer science, empirical aesthetics and art history. He lives in Wales (and in his studio).

Made in United States
North Haven, CT
14 January 2025

64420898R00124